Maker Innovations Series

Jump start your path to discovery with the Apress Maker Innovations series! From the basics of electricity and components through to the most advanced options in robotics and Machine Learning, you'll forge a path to building ingenious hardware and controlling it with cutting-edge software. All while gaining new skills and experience with common toolsets you can take to new projects or even into a whole new career.

The Apress Maker Innovations series offers projects-based learning, while keeping theory and best processes front and center. So you get hands-on experience while also learning the terms of the trade and how entrepreneurs, inventors, and engineers think through creating and executing hardware projects. You can learn to design circuits, program AI, create IoT systems for your home or even city, and so much more!

Whether you're a beginning hobbyist or a seasoned entrepreneur working out of your basement or garage, you'll scale up your skillset to become a hardware design and engineering pro. And often using low-cost and open-source software such as the Raspberry Pi, Arduino, PIC microcontroller, and Robot Operating System (ROS). Programmers and software engineers have great opportunities to learn, too, as many projects and control environments are based in popular languages and operating systems, such as Python and Linux.

If you want to build a robot, set up a smart home, tackle assembling a weather-ready meteorology system, or create a brand-new circuit using breadboards and circuit design software, this series has all that and more! Written by creative and seasoned Makers, every book in the series tackles both tested and leading-edge approaches and technologies for bringing your visions and projects to life.

More information about this series at https://link.springer.com/bookseries/17311.

Learning Algorithms for Internet of Things

Applying Python Tools to Improve Data Collection Use for System Performance

G.R. Kanagachidambaresan
N. Bharathi

Apress®

Learning Algorithms for Internet of Things: Applying Python Tools to Improve Data Collection Use for System Performance

G.R. Kanagachidambaresan
Vel Tech Dr. RR & Dr. SR Technical Unive
Chennai, Tamil Nadu, India

N. Bharathi
Chennai, Tamil Nadu, India

ISBN-13 (pbk): 979-8-8688-0529-5
https://doi.org/10.1007/979-8-8688-0530-1

ISBN-13 (electronic): 979-8-8688-0530-1

Managing Director, Apress Media LLC: Welmoed Spahr
Acquisitions Editor: Susan McDermott
Desk Editor: James Markham
Editorial Project Manager: Jessica Vakili

Distributed to the book trade worldwide by Springer Science+Business Media New York, 1 New York Plaza, Suite 4600, New York, NY 10004-1562, USA. Phone 1-800-SPRINGER, fax (201) 348-4505, e-mail orders-ny@springer-sbm.com, or visit www.springeronline.com. Apress Media, LLC is a California LLC and the sole member (owner) is Springer Science + Business Media Finance Inc (SSBM Finance Inc). SSBM Finance Inc is a **Delaware** corporation.

For information on translations, please e-mail booktranslations@springernature.com; for reprint, paperback, or audio rights, please e-mail bookpermissions@springernature.com.

Apress titles may be purchased in bulk for academic, corporate, or promotional use. eBook versions and licenses are also available for most titles. For more information, reference our Print and eBook Bulk Sales web page at http://www.apress.com/bulk-sales.

Any source code or other supplementary material referenced by the author in this book is available to readers on GitHub. For more detailed information, please visit https://www.apress.com/gp/services/source-code.

If disposing of this product, please recycle the paper

To my family members, Dr. Mahima V, sons Ananthajith K and Anuthaman K, and daughter Anugraha K; students; scholars; and dear friends.

—G.R. Kanagachidambaresan

To my family members, scholars, students, and dear friends.

—N. Bharathi

Table of Contents

About the Authors

G.R. Kanagachidambaresan is a professor in the Department of Computer Science and Engineering at Vel Tech Rangarajan Dr. Sagunthala R&D Institute of Science and Technology. He completed his PhD in 2017 at Anna University and currently handles funded projects from ISRO, DBT, and DRDO. He has written books and articles on the topics of IoT, wireless networks, and expert systems. He has also consulted for leading MNC companies. He is the managing director for Eazythings Technology Private Limited and a TEC committee member for DBT. He is also the editor in chief for the Next Generation Computing and Communication Engineering series from Wiley.

N. Bharathi is an associate professor in the Department of Computer Science Engineering at the SRM Institute of Science and Technology. She has previously been an associate professor at the Saveetha School of Engineering, the R&D head at Yalamanchili Manufacturing Private Limited, and an assistant professor at SASTRA Deemed University. She has working knowledge of the Internet of Things and embedded systems as well as the cloud computing and big data domains. She was awarded a PhD degree in computer science in 2014 from SASTRA Deemed University, having 20 years of work experience as an academician and one-and-a-half years of industrial experience on the ARM platform with Ubuntu OS. She has published many research papers in reputed peer-reviewed journals and conferences and books and guided students in various domains of computer science engineering. She is currently guiding five research scholars in cloud, big data, IoV, and other advanced domains. She is a life member of ISTE, senior member of IEEE, and AICTE certified master trainer on high-performance computing.

About the Technical Reviewer

Massimo Nardone has three decades of experience in security, web/mobile development, and cloud and IT/OT/IoT architecture. His true passions are security and Android. He has been programming and teaching how to program with Android, Perl, PHP, Java, VB, Python, C/C++, and MySQL for more than 30 years. He holds a master's degree in computing science from the University of Salerno, Italy. He has worked as a chief information security officer (CISO), software engineer, chief security architect, security executive, OT/IoT/IIoT security leader, and security architect for many years. He is currently the VP of OT security for SSH Communications Security.

Acknowledgments

Our sincere thanks to Jessica Vakili, Shobana Srinivasan, and the entire Apress team. Our sincere thanks to Vel Tech and SRM management and DBT – INDIA (BT/PR47509/AAQ/3/1058/2022).

Preface

The way we connect with our surroundings, gadgets, and systems has completely changed as a result of the Internet of Things' (IoT's) exponential rise. An enormous quantity of data is being generated every day due to the widespread use of smart devices and sensors, providing previously unheard-of opportunities to improve efficiency, automation, and decision-making in a variety of fields. But rather than stopping at data collection and conveyance, the real promise of IoT is in turning that data into insights that can be put to use. This is when learning algorithms—which include methods from both deep learning and machine learning—come into play.

To construct intelligent systems that can self-optimize, foresee the future, and adapt to changing environments, this book explores the integration of learning algorithms with the Internet of Things. We can create smarter systems that tackle a variety of societal issues, from transportation and smart cities to healthcare and agriculture, by combining the advantages of IoT with machine learning. The deliberate use of learning algorithms makes the idea of "smartness" more than just a catchphrase; it becomes an actual reality.

This book offers readers a thorough understanding of how learning algorithms can improve real-time IoT applications. With its coverage of fundamental ideas, useful applications, and real-world scenarios, it offers a comprehensive method for utilizing these technologies to boost system efficiency and data use. Code samples and thorough descriptions showing how to apply these techniques for different applications are especially helpful for researchers and developers.

Funding Information

Part of this book is supported by the Indias Department of Biotechnology (BT/PR47509/AAQ/3/1058/2022).

CHAPTER 1

Introduction to Learning Algorithms

Learning algorithms are capable of extracting features from the input data and gaining intelligence to identify and predict new input data. This is like a human learns from his birth. The input data has the question as well as the answer, such as an image of an object and its name. How a baby learns the names of objects as they grow up is similar to how computer programs implemented with mathematical and logical computations are used as learning algorithms.

Generally, the learning algorithms can be machine learning algorithms, deep learning algorithms, genetic algorithms, and supporting optimizers. The commonality behind all the learning algorithms is that they extract information from the input training data and apply the gained knowledge to make predictions and identify new input data.

The machine learning algorithms enable the computers to gain knowledge from the input data automatically. The past data fed as input is used to train the mathematical models in order to predict the future data. The building blocks of deep learning algorithms are artificial neural networks, which form the basis for computation and learn the features of the data. As the number of layers increases appropriately in an artificial neural network, the accuracy will increase, and the algorithm will learn the features with fewer resources.

© G.R. Kanagachidambaresan and N. Bharathi 2024
G.R. Kanagachidambaresan and N. Bharathi, *Learning Algorithms for Internet of Things*,
Maker Innovations Series, https://doi.org/10.1007/979-8-8688-0530-1_1

Genetic algorithms are a class of evolutionary algorithm that simulates a genetic framework and natural selection. Various species evolve and survive across generations by reproducing. Similarly, the problem and solution search space is shared by individual data points based on a fitness function. As a generation evolves, the population converges to a best solution. The optimizers support the learning process by finding the best set of attributes or modify the attributes involved in the neural networks such as the learning rate and weight to minimize certain losses due to errors in the data.

1.1 Genetic Algorithms

A genetic algorithm (GA) is based on biological evolution and natural selection. These are heuristic search-based algorithms that can be used in machine learning, artificial intelligence, and optimization techniques. Generally, these algorithms are useful for real-time applications that are complex in nature and require longer time to resolve such as image processing, circuit design in electronics, etc.

1.1.1 Identifying Key Terms

Some basic key terms are necessary to understand when discussing genetic algorithms such as *selection*, *crossover*, and *mutation*. Each generation of the population possesses components such as *genes*, *chromosomes*, *alleles*, *fitness functions*, etc. Also, each generation of the population is a subset of all possible solution spaces that have solutions for the given problem.

- **Chromosome** represents a single specific solution in the population.

- **Gene** is an element in the chromosome that contributes to the specific solution and is also called a decision variable.

- **Allele** is the value given to a gene that belongs to a specific chromosome.

- **Fitness function** determines the suitability of the solution to resolve the problem from the specific solution as input along with the other attributes required. The output of the fitness function is also called the *fitness score*. Genetic operators are dependent on the fitness score.

- **Selection** is one of the genetic operators that helps to choose individuals from the current population with a better fitness score than others to produce the next generation.

- **Crossover** is another operator that identifies the crossover region in the gene of the two individuals and exchanges those regions to generate a new individual (offspring).

- **Mutation** is the third operator, which inserts random genes of the offspring generated through crossover in order to prevent the premature convergence and preserve the diversity across the generations.

Genetic operators enable genetic algorithms to solve problems in a better way than random search algorithms because genetic algorithms use historical data to identify and narrow down the best solution regions in the solution space. The operators manipulate the chromosomes and its genetic structures as each chromosome contribute to a set of possible solutions. Also, along with the genetic operators, fitness function, and its score, the genetic algorithm is generating the new generations to move closer to the best solution space.

1.1.2 Generation Stages

A genetic algorithm works based on the evolutionary generation cycle of the population to yield better solutions. The following are the various stages through which each generation is produced:

- **Stage 1:** The first stage is to initialize the population. The initial population must have all the possible solutions of the problem given. The random binary strings method is one of the famous methods to generate the initial population. It is set to either 0 or 1 for the gene, which represents a variable of the problem, and genes are combined to form a chromosome, which is a single instance/individual of the population and a single solution.

- **Stage 2:** In the second stage, the fitness function is applied on the population to compute the fitness score of every individual. The fitness score is denoted by a numerical value, which is based on the fitness function and the attributes of the problem. The individuals that have higher fitness scores gain the chance of existence to generate the next generation.

- **Stage 3:** The third stage is the selection, which chooses the individuals for reproduction to produce the next-generation offspring. The chosen individuals are paired with another individual to improve the reproduction. Their genes are used to proceed further with the fitness-based selection technique.

- **Stage 4:** In the fourth stage, two key operators, crossover and mutation, are used on each selected pair in the third stage. This stage is responsible for generating the next-generation population, called *reproduction*. The two individuals paired are taken, and their genes are selected randomly to generate the offspring. The same process is repeated on every pair of individuals to generate the new child generation. Next, mutation adds new useful genetic information to the individuals of the child population by flipping certain bits in the binary string representing the chromosome, as illustrated in Figure 1-1. This fixes the local minimum problem and enables diversification.

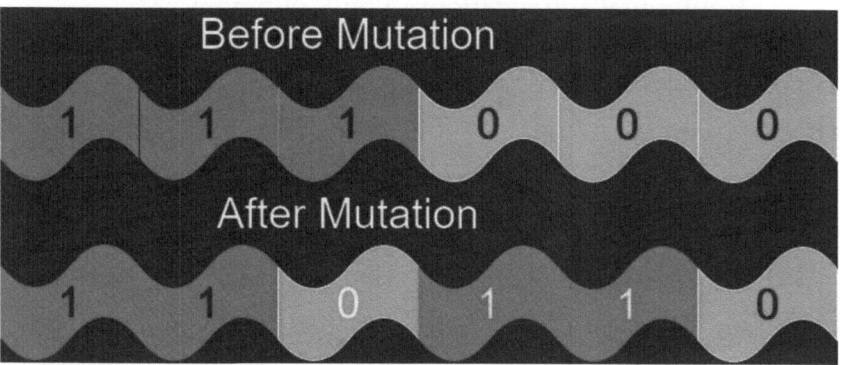

Figure 1-1. *Mutation process*

- **Stage 5:** The fifth stage is the process of replacing the less fit individuals from the old population with the higher fit individuals generated from the new child population. This results in better-quality solutions produced for the given problem.

- **Stage 6:** In the sixth stage, the fitness score of the new individuals is compared with the threshold fitness score. If it is reached, then the algorithm terminates; otherwise, it proceeds with generating the next new generation and continues through all the previous stages until it identifies the best solution.

1.1.3 Generation Applications

The applications that apply genetic algorithms are as follows:

- **DNA analysis:** The sequence of DNA is analyzed for criminal investigation, medical diagnosis, etc.

- **Multimodal optimization:** Multiple optimum solutions exist for certain problems; they are really required to handle applications that are diverse in nature.

- **Transportation and telecommunication routing:** The product delivery to customers follows an optimized route, which reduces the cost of travel and time spent.

- **Scheduling:** Various problems like fixing timetables, determining dates of appointments, etc., are well solved by GA.

- **Economics:** Models such as cobweb, game theory, asset valuing, etc., are having diverse set of solutions, and the best one can be determined efficiently with genetic algorithms (GAs).

- **Aircraft design:** The design of an aircraft is based on the parameters that can be manipulated to generate better designs.

- **Robotics and automotive design:** The movement of robots and the design of appearances of motor vehicles have multiple possible solutions, and the GA is applied to determine the best solution using the required parameters.

In conclusion, the genetic algorithms are well suited for applications that inherently require parallel processing and the optimizing of performance from multiple objective problems, and it does not depend on the derivative information. The pitfalls of GAs are as follows: they are over-suitable for simple problems, have no guaranteed final solution, consume more computation time for certain applications, and lead to suboptimal solutions if the parameters are not mapped properly to the fitness function.

1.2 Machine Learning

Humans learn from their past experiences, and they adapt accordingly to the environment. Alternatively, machines have simply worked based on their instructions or mechanical configurations. However, more recently, researchers have asked the question, why can't machines learn? The answer to this question is machine learning, with which machines can gain knowledge from the set of inputs from the past and predict the future.

Machine learning is a subsection of artificial intelligence mainly for enabling machines to gain information from the existing data on their own and act accordingly with future data such as email filtering, recommendation systems, auto-tagging, etc.

The term *machine learning* was coined in 1959 by a famous researcher Arthur Samuel who contributed many works to the field of artificial intelligence. The general definition of machine learning is "Machine learning enables a machine to automatically learn from data, improve performance from experiences, and predict things without being explicitly programmed."

7

In machine learning, the predictive models are built by combining the concepts of computer science and statistics. The predictive models are also called mathematical models, and they influence decisions or predictions by learning from historical data or training data. The learning process is automated using the machine learning algorithms and their implementations. Once the learning with all its input is done, the model starts predictions whenever the new data arrives. The performance or accuracy of predictions could be increased based on the quantity and quality of historical data or information provided.

Machine learning has transformed our perception of how we look at problems. Developers need to focus on how input is fed to the learning algorithms, instead of how the predictions need to be performed in the case of real-world problems. Based on the algorithms, the mathematical model is built on the machine. The model itself formulates the logic as it starts receiving input data for learning.

The necessity of machine learning is growing rapidly as the process of accomplishing certain complex tasks is very time-consuming for a developer or programmer to code directly. For humans, envisioning one, two, and three dimensions is easy, but the dimensions beyond than that and the manipulations on them can be solved only using complex algorithms and equations. Machine learning can use data to identify patterns and study its features in the input dataset. It gets easier as more and more data arrives; the machine learns constantly and adapts automatically based on its learning and improves its predictions. The data is the key factor in machine learning, and hence it is a data-driven technology. Overall, machine learning techniques save time, effort, and money.

Generally, for all methodologies and technologies, measuring performance is important in order to study how well they are designed and developed. In that way, cost plays the key role in machine learning for analyzing the performance of machine learning algorithms. It estimates the relationship between the input data and output prediction.

It computes the difference between the actual output and the predicted output and aggregates it as a single value. The popular machine learning algorithms are logistic regression, linear regression, naïve Bayes, decision tree, random forest, K-nearest neighbor (KNN) classification, support vector machine (SVM), and gradient boosting. These are categorized under supervised machine learning. K-means, apriori, hierarchal clustering, anomaly detection, principal component analysis, and independent component analysis are unsupervised machine learning algorithms.

Machine learning is used in diverse domains such as face or fingerprint recognition, product recommendations in online stores, friend suggestions on Facebook, and profile recommendations on LinkedIn. Many top companies employ machine learning techniques to study customers' interest using a large volume of data and recommend suggestions about products. The following are more use cases:

- **Gmail's spam filter:** Gmail keeps track of malicious links and maintains a large database to filter the junk incoming emails. The spam filter also trained and learned to filter mails based on suspicious words in the subject of emails. It uses machine learning algorithms and natural language processing techniques to classify emails into spam or not.

- **Face recognition in Facebook:** The DeepMind face verification system used by Facebook is built based on machine learning and deep learning algorithms. Auto tagging of faces and detection of existence of one's face in other albums are practiced in Facebook. But currently, Facebook restricts the face recognition system to verify identity on personal devices and in financial services.

- **Netflix's movie and show recommendations:** The recommendation engine at the core of Netflix suggests the entertainment programs, and more than 75% of suggestions are followed by viewers. Machine learning algorithms are used in the recommendation engine.

- **Amazon's diverse usage:** Product recommendations, supply chain management, audio and video recognition, and fraud detection are used at Amazon. For example, Amazon's Alexa uses machine learning and natural language processing to process and perform user commands such as playing specific songs, booking a cab, connecting other home appliances and health gadgets, etc.

The amount of data generated every second is increasing every day. Predictive models can be used to study the nature of data and extract useful insights for decision-making. Many top-level companies use machine learning models and generated data to overcome unwanted risks and make decisions to improve profit. The importance of machine learning is very clear as it analyzes the increased generation of data, revealing patterns or trends in the data, cracking complex problems, and making decisions.

1.3 Deep Learning

Deep learning techniques work well with the large datasets. The layered approach of deep learning discovers hidden features. To solve real-world problems, deep learning algorithms need to be studied thoroughly. The best solution can be easily identified with the right understanding of deep learning algorithms. The knowledge of deep learning is necessary and highly valuable for researchers and professionals. As all the domains are utilizing the deep learning techniques for advancement and improvement, deep learning enables the professionals to be competent enough and develop state-of-the-art applications. Also, deep learning techniques are continuously emerging, and professionals need to be up-to-date to gain knowledge constantly.

Deep learning follows representation learning and is a subcategory of machine learning. High-level complex representations of features are learned automatically by machines from the dataset. Challenging tasks such as image recognition, speech recognition, natural language processing, etc., are handled efficiently with deep learning algorithms, and the models generated better results.

Artificial neural networks (ANN) are used in deep learning algorithms in order to learn the complex features from the dataset with the increased number of layers in the network. ANNs consist of artificial neurons that form a network to work together and mimic the biological neural network functioning. An ANN comprises three layers: an input layer, one or more hidden layers, and an output layer. Every layer comprises of artificial neurons called *nodes*, which are connected to achieve complex tasks. Each node is a computation unit that takes one or more input and produces output.

The independent variables values are fed through the input layer into the ANN. The hidden layers are meant for learning and understanding the independent variables, the relationships among them, and their contribution in determining the dependent variable values. The output layers are responsible for producing the output results based on the processing happening in the hidden layers. The number of hidden layers is the key factor in deciding whether the neural network is a shallow or deep network. As the number of hidden layers increases, the network changes from shallow to deep and produces more and more accurate results.

Shallow neural networks are for simple networks that implement classification, regression, etc. Generally shallow neural networks have one or two hidden layers. Deep neural networks consist of a greater number of hidden layers, and all neurons in each hidden layer learn to extract specific features from the dataset. Each added hidden layers is enabling the learning of complex datasets and its patterns. Finally, the output layer combines the results produced by the hidden layers based on the requirement of the task, as shown in Figure 1-2. Though the deep neural networks are computationally intensive, they give more accurate results than shallow neural networks. Hence, deep neural networks are most desirable for applications involved in real-time activities such as natural language processing, image and video processing, and multivariate time series-based and real-time forecasting applications.

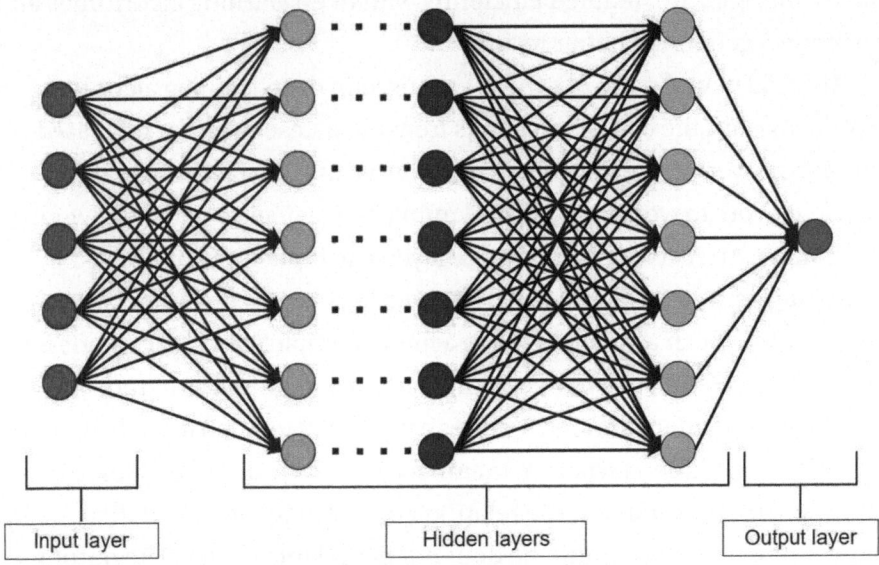

Figure 1-2. *Deep neural network model*

Shallow as well as deep neural networks work by accepting input data or previous layer data, applying computations on received data based on the activation function assigned for that layer, and detecting the pattern. Consequently, the weights are assigned to the data at neurons, which contribute to the further processing and prediction as the data moves toward the output layer. The output layer generates a label for each class of data or predicts any other values, etc., based on the combination of hidden layer outputs.

The two different models based on the flow of information in the deep neural network are forward propagation and backpropagation. In the former model, the input signal moves toward the output layer through hidden layers one by one. As the name of the model, the signal moves only in the forward direction. In back propagation, the second model uses the chain rule to examine the involvement of each neuron to generate the output as well as errors. Later the error value is fed back to the network, and the adjustment in the weights of the neuron is performed to reduce the errors in the output. In addition, there exists optimization techniques that also contribute to the reduction of errors in back propagation models. You can find a detailed discussion of optimization techniques later in this chapter.

Apart from the two models, two more functions used in deep neural network play key roles. They are the activation function and the loss function. The activation function converts the values of input data to a form that can be understandable by the deep neural network. For example, the sigmoid activation function, which is an "S" shaped curve takes any real number as input and produces output in the range of 0 to 1. Similarly, other activation functions also exist such as tanh, linear, etc. A loss function is a performance measure that indicates the how well a neural network produces output. Mean absolute error, mean square error, cross entropy, accuracy, etc., are some of the loss functions that compute the loss or difference in the actual and predicted values using deep neural networks and deep learning algorithms.

The algorithms that are most common in deep learning are convolutional neural networks (CNNs), recurrent neural networks (RNNs), and long short-term memory networks (LSTMs). In addition, generative adversarial networks (GANs), multilayer perceptron (MLPs), radial basis function networks (RBFNs), deep belief networks (DBNs), restricted Boltzmann machines (RBMs), autoencoders, self-organizing maps (SOMs), and transfer learning algorithms are also used in specific applications of deep learning.

The applications that are using the previous deep learning algorithms are as follows: CNNs perform well when processing satellite images and detecting anomalies. RNNs are used for natural language processing and handwritten character recognition. LSTMs are good at time-series prediction and music composition. GANs support rendering 3D objects and creating cartoon characters. MLPs are the best choice for speech recognition and machine translation. RBFNs are good at classification and regression. DBNs are better in video recognition and motion capturing. RBMs are used in collaborative filtering and feature learning. Auto encoders are well suited for pharmaceutical discovery and popularity prediction. SOMs help in understanding the high-dimensional data with visualization.

1.4 Optimization

Optimization is generally a mathematical technique that has an objective function to be either maximized or minimized based on the nature of the problem in order to resolve. The properties of the variables involved in the problem determine the type of approach used to solve. In the machine learning and deep learning context, the optimizers are algorithms that are used to modify the learning rates and weights assigned in the neural network in order to reduce the loss or improve the results. Hence, optimizers are used to solve minimized optimization problems. The optimization is done by comparing the results of every iteration by changing the parameters involved in the model until the optimum results are generated. The basic terms you need to understand at this juncture are *sample*, *epoch*, and *batch*. *Sample* is a single entry or row in the given dataset. *Epoch* denotes the number of iterations the algorithm executes on the entire training dataset. *Batch* represents the number of samples to be chosen for updating the attributes or parameters of the model.

There are various algorithms for optimizing the model and its results. Two main optimizers that are the simple and oldest techniques are called *gradient descent* and *stochastic gradient descent*. Before discussing the optimizers, the fundamental terms that everyone should be aware are *global maxima, global minima, local maxima,* and *local minima*. In a mathematical function, the smallest value within a given range, and not the entire series, is called the *local minima*, and the largest value in each range of the function is called the *local maxima*. Alternatively, the smallest value over the entire series of the function is called the *global minima* and that of the largest is the *global maxima*. Generally, more than one local minimum and local maxima are possible, but there exists only one global minima and maxima, as in Figure 1-3.

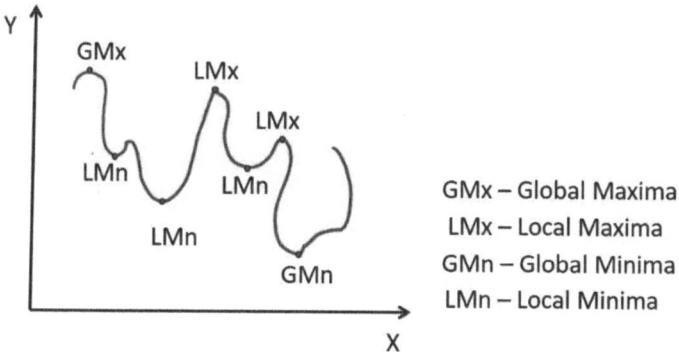

Figure 1-3. *Global/local maxima/minima*

Finally, another important term is the learning rate, which denotes how fast the neural network reaches the optimum. It determines the step size in each iteration to move toward the minima of the mathematical function. But the learning rate needs to be decided with utmost care because if it is a low value, then the model becomes overfit, and if it is high value, then it ends up in the underfit of the model. Also, we should not keep it as constant through the learning process, as it leads to an oscillation problem of swinging between two values alternatively forever and never reaching minima, as illustrated in Figure 1-4. Hence, the learning rate needs to be reduced as moving forward the number of iterations.

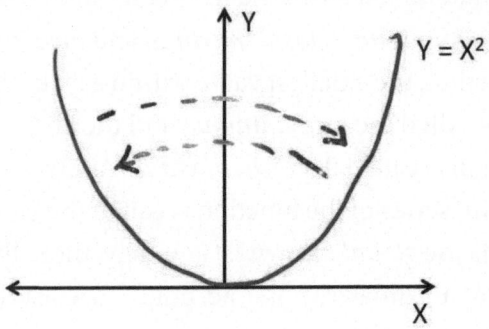

Figure 1-4. *Oscillation problem*

1.4.1 Types of Optimizers

There are many types of optimizers based on computing gradients and applying momentum.

- **Gradient descent (GD) optimizer:** This is a first-order iterative algorithm that repeats the process of finding global/local minima/maxima, each time in opposing direction of the gradient. The gradient represents the slope along the direction of change of any scalar function. This algorithm works for functions that must be differentiable and convex. There are three types of gradient descent algorithms: batch gradient descent, stochastic gradient descent, and mini-batch gradient descent. The types are categorized based on the time complexity of the algorithm. When large data points are involved, it is $O(RN^2)$, where R is the number of iterations and N is the number of data points.

- **Batch gradient descent optimizer:** This is a derivative of the gradient descent optimizer, which uses the entire dataset for computing the gradient and other attributes required for each iteration. It computes the error for every training sample for each epoch. Though this leads to stable error gradient and convergence, it requires the entire training dataset to exist in memory. Also, it becomes slow if the dataset is large.

- **Stochastic gradient descent optimizer:** This is different from the batch gradient descent optimizer in the way it updates the parameter one at a time and not the entire dataset. For example, if the model contains a 1K dataset gradient, it will update the model parameters 1,000 times one at a time. Hence, this optimizer requires less memory compared to the batch gradient. But in certain models, frequent updates will result in noisy gradients and in turn increase the error.

- **Mini-batch gradient descent optimizer:** This combines the benefits of both batch GD and stochastic GD. It creates a number of small batches from the training dataset and updates each batch one at a time. Hence, this method provides the robustness of stochastic GD and efficiency of batch GD. The batches are chosen at random, ranging from 50 to 250 samples.

- **Adagrad:** This changes the learning rate for each epoch, depending on the other parameters while training. The gradient adapts the change in the parameters involved in the model. Hence, it's called *adaptive gradient descent* (Adagrad). The change in the learning rate is inversely proportional to the change in the parameters. This approach is useful when subjected to real-world datasets, which are sometimes sparse/dense in nature. Also, in this method there is no need to change the learning rate manually.

- **RMSProp:** The root mean square propagation (RMSProp) optimizer is a popular optimizer that uses the difference in the sign of the two gradients. If the sign is the same, the direction is right, and the step size is increased. If the sign is different, then the step size must be decreased. After determining the step size, the weights are updated. This optimizer converges soon with less tuning but at the cost of determining the learning rate manually.

- **Adadelta:** Adadelta is a variant of Adagrad, which overcomes the drawback of specifying the learning rate manually. Also, the learning rate becomes smaller as you move toward iterations, and from certain later iterations it will no longer learn anything. Also, the learning rate is computed by considering only the past w gradients instead of considering all the past gradients, where the w is the window size. For example, if w = 15, only the past 15 gradient values are squared in determining the current learning rate. This ensures that the learning rate at later iterations may not be very low.

- **Momentum:** This is similar to concept of a ball rolling on a slope gaining momentum as it rolls down. In this method, the fraction of previous update of weights is added to the current weight update. Instead of directly updating the weights, this method introduces a new parameter called *momentum*, which signifies the moving gradient. The graph of learning shows larger steps in the horizontal direction (faster learning) and smaller steps in the vertical direction (slower learning) during updating and signifies that learning takes place with more momentum as it moves toward the goal.

- **Nesterov momentum:** This is a modified version of momentum. In this method, instead of computing the cost function before the weight updating, the reverse happens. The weights are updated based on gradient and momentum, and then the gradient of the cost function is computed for next location. This leads to the more accurate gradient computation and converges faster than the momentum optimizer.

- **Adam:** The name of this optimizer is derived from adaptive moment estimation. It is a variant of stochastic gradient descent (SGD) as it is not using a single learning rate. Also, it inherits the characteristics of Adagrad and RMSProp as they are extensions of SGD. RMSProp computes the learning rate based on the first moment (mean), whereas the adam optimizer computes the learning rate based on the second moment.

- **Adamax:** This is a variant of Adam, which is based on infinity norm (measures how large the vector is by the magnitude of its largest entry). It is suitable for time variant process such as speech or music whose noise conditions change dynamically. The optimizer is stable enough as it is based on infinite order and involves only less tuning of parameters.

- **SMORMS3:** Squared mean over root mean squared cubed (SMORMS3) is a variant of RMSProp. So unlike RMSProp computes a simple average of squared gradient, it is calculating the cube root of the moving average of the cube of the squared gradients. Similar to RMSProp, it includes a damping factor (ensures that the learning rate is proportional to the inverse square root of the variance of the gradients) to prevent a learning rate that is too big. It is useful in an application where a high variance in the gradient exists.

1.5 Summary

This chapter provided fundamental knowledge of learning algorithms. With this basic knowledge, anyone can proceed with programming the learning algorithms. Programming learning algorithms involves the calling of methods, which implement the algorithms already for any given dataset that can be passed as parameters along with other hyper parameters required for the algorithm. In the next chapter, the packages that offer the learning algorithm methods will be discussed.

CHAPTER 2

Python Packages for Learning Algorithms

Learning algorithms are widely used in almost all domains. The overall procedure to generate a learning algorithm model is to preprocess the dataset, train the model based on the nature of the data with supervised or unsupervised or deep learning algorithms, and then verify the model by testing. These procedure steps are provided in many Python packages. Keras, TensorFlow, SciPy, PyTorch, Theano, Pandas, Matplotlib, Scikit-learn, Seaborn, and OpenCV are the 10 most important Python packages that support learning algorithms, their preprocessing and output prediction, the visualization of results, etc. This chapter describes these 10 packages.

2.1 Keras

Keras is an open-source software library that serves as an application programming interface (API) to the TensorFlow platform. It is an easy-to-use API that simplifies a developer's work by providing the building blocks for machine learning and deep learning implementations.

The word *Keras* has its origin from the green word (κέρας), which means horn. Keras was developed mainly for research work on the Open-ended Neuro-Electronic Intelligent Robot Operating System (ONEIROS). Now the Keras API is used for developing software in many well-known companies such as YouTube, Twitter, NASA, Waymo, etc.

© G.R. Kanagachidambaresan and N. Bharathi 2024
G.R. Kanagachidambaresan and N. Bharathi, *Learning Algorithms for Internet of Things,
Maker Innovations Series*, https://doi.org/10.1007/979-8-8688-0530-1_2

2.1.1 Features of Keras

The following are some of the key features of Keras:

- **Ease to use:** Keras is a simple and steady API that helps users, developers, and programmers with general requirements and use cases. It also provides easily understandable error messages. For advanced use cases, it offers a clear step-by-step path to develop complex models.

- **Faster:** It offers experimental iterations and cycles, so debugging is achieved in a rapid manner. XLA compilation and autograph optimizations support speediness in the execution of models. The code's conciseness and maintainability also contribute to the quicker execution.

- **Readily scalable:** The foundation of Keras is built in a way that can scale to huge GPU clusters and even one complete TPU pod. Since Keras models are built on top of TensorFlow, the scalability is easier for most of the industry-oriented software infrastructures.

- **Flexible:** Keras has a gamut of features from low-level implementation features to high-level features to realize the research ideas. It offers rich collections of APIs and built-in models to implement any kind of application.

- **Wide support:** Keras utilizes the full deployment feature of the TensorFlow platform to use Keras models in a range of application platforms such as utilities in browsers, apps in iOS and Android, web APIs, apps in embedded devices, etc.

- **Rich documentation:** Keras insists developers create guides, which improves the wide usage of the library APIs.

2.1.2 Keras Ecosystem

Keras, in addition to its efficient deep learning API, supports an ecosystem
for a wide range of AI, machine learning, and deep learning frameworks in
various environments. This ecosystem makes it easy to use and optimizes
the training and deployment of models. The ecosystem, as illustrated
in Figure 2-1, covers the natural language processing, computer vision,
machine learning models on browsers and mobile phones, determination
of best hyperparameters for the learning model, management of the
production pipeline for deep learning models, and the optimization toolkit
for deploying ML models.

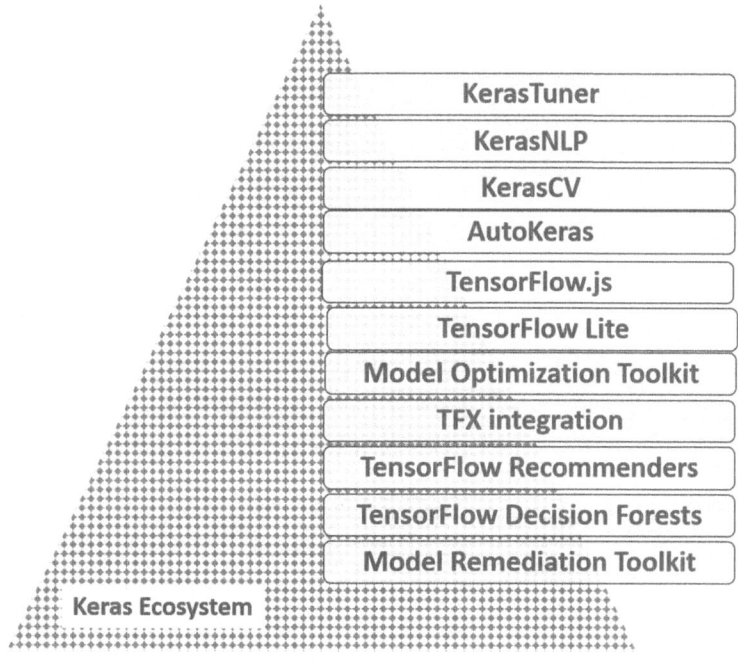

Figure 2-1. *Keras Ecosystem*

The following offers a breakdown of the components:

Keras Ecosystem Component	Description
KerasTuner	This framework mainly focuses on the hyperparameters, which determine when the learning algorithms ends up with the training. It helps to decide the best hyperparameter for any model.
KerasNLP	Keras NLP is tailored to the natural language processing (NLP). This offers a complete NLP framework to support the development of NLP-based applications with built-in models with a customization option.
KerasCV	The KerasCV framework is mainly for computer vision applications. The modular components of KerasCV have Keras core as their base and also work with TensorFlow, PyTorch, and JAX.
AutoKeras	This framework makes the machine learning algorithms accessible easily by everyone working in applications of different domains. It readily supports image/text classification, regression, etc.
TensorFlow.js	Tensorflow.js is mainly focused on the browser-based deployment of deep learning models. The pretrained models can be imported even on the web browser for retraining. With knowledge of JavaScript alone, users new to machine learning can use it.

(continued)

Keras Ecosystem Component	Description
TensorFlow Lite	This is a framework tailored for supporting deep learning models on embedded/mobile devices. It converts the trained models into smaller optimized versions in the form of a `.tflite` file, which is suitable in mobile devices.
Model Optimization Toolkit	The models of machine learning in mobile and embedded devices are optimized for latency cost, storage cost, and message cost in wide area network devices using this framework. This is quite useful where the computing and storage resources are limited.
TFX integration	TensorFlow Extended (TFX) is a framework for configuring the components to control and monitor the deep learning models and systems. IT focuses on creation and management of deep learning models deployed in a production pipeline.
TensorFlow Recommenders (TFRS)	This framework is tailored for building recommendation systems with a steady learning curve even with dynamically changing behaviors of users. Also, it is more flexible to build large complex models.
TensorFlow Decision Forests (TF-DF)	TF-DF spotlights on training and interpreting decision forests (random forests, gradient boosted trees) models. It supports regression, classification, ranking, and uplifting models.
Model Remediation Toolkit	This framework mainly focuses on resolving the fairness and biasing in the models of deep learning and machine learning and helps to work with objectivity.

2.1.3 Layers in Keras

The layers are as follows:

Keras Layer	Functionality
The base Layer class	This forms the base from which all other layers are developed by inheriting it. A layer is basically a callable object that consumes tensors as input and produces tensors as output.
Layer activations	This governs the usage of activations. The activation function is applied on the layers or through the argument in layer methods. Many built-in activations exist as string identifiers to pass as values of the activation argument.
Layer weight initializers	This initializes the weights of Keras layers with the help of keyword arguments such as `kernel_initialzer` and `bias_initializer`. Built-in initializers can also be passed as string identifier. In addition, custom initializers can be created if required.
Layer weight regularizers	This applies penalty on layer activity or parameters during learning optimization. These penalties are contributing for the loss function that in turn is optimized by the network. To regularize, three main keyword arguments are `kernel_regularizer`, `bias_regularizer`, and `activity_regularizer`.
Layer weight constraints	To apply constraints on the weight of each layer, the `tf.keras.constraints` module defines the model parameters or keyword arguments to set constraints during the learning process.

(continued)

Keras Layer	Functionality
Core layers	This layer has the following components: input object (initialize tensor), dense layer (densely connected neural network layer), activation layer (applies activation function to the output), embedding layer (converts +ve integers to fixed size vectors), masking layer (masks sequence to skip timesteps), and lambda layer (wraps arbitrary expressions for quick prototype). These components form the core layer.
Convolution layers	The convolution kernel can be created with the layer input over the required dimensions such as 1D, 2D, and 3D to produce output tensors.
Pooling layers	It is for pooling operation on 1D, 2D, and 3D layers. It supports Max, Average, GlobalMax, and GlobalAverage pooling.
Recurrent layers	Recurrent neural networks (RNNs) are enabled using this layer. It supports long short-term memory (LSTM), gated recurrent unit (GRU), simple RNN, time distributed, bidirectional, convolutional LSTM, and base RNN.
Preprocessing layers	The preprocessing layer supports the building of a Keras-based input processing pipeline. These pipelines serve as an independent preprocessing code block that can be used with Keras models. Later the combination is used as saved models. Text, image, and feature preprocessing methods are available.
Normalization layers	Normalization is to convert all the data into one type and within a range as per the requirement. Batch normalization is applied for transforming the data in such a way that its mean is close to 0 and standard deviation is near to 1.

(continued)

Keras Layer	Functionality
Regularization layers	This layer sets the dropout in the input randomly to 0 during training to avoid an overfitting possibility. Also, this layer has methods to update the cost function with respect to input activity.
Attention layers	This layer is responsible for focusing selectively on important portions of input. This layer prioritizes and emphasizes necessary information to improve the performance of the model.
Reshaping layers	The shape of the input is modified based on the requirement. It is a layer that enables the alteration of structure of the model layer without affecting data.
Merging layers	This layer merges more than one input tensors of same shape into a single tensor of the same shape. This is mainly useful in deep learning algorithms where fusion of multiple images, text, or input from various sensors.
Activation layers	This layer is responsible for providing various activation function layers such as rectified linear unit (ReLU), Softmax, parametric ReLU, exponential and thresholded activation function.

2.2 TensorFlow

TensorFlow is an open-source software library framework especially developed for machine learning and deep learning algorithms. It was created by the Google team to analyze, design, and develop the ideas and concepts in artificial intelligence, machine learning, and deep learning.

The basic building blocks of TensorFlow are tensors. A tensor is an N-dimensional data type ranging from zero to multidimensional arrays of numbers. Generally, tensors are of any dimension and contain data with identical data types. Algorithms and mathematical computations are implemented as graphs with vertices as tensors, and edges are connections between the tensors, which determines the input and output associations.

2.2.1 Features of TensorFlow

The following are the key features of TensorFlow:

- **Simple to implement:** TensorFlow offers various features and functions at multiple levels. As more and more functions are available, the opportunity of choosing the best to easily build models for machine learning and deep learning algorithms is highly possible.

- **Flexible:** Instant iteration, flexible execution, and spontaneous debugging are supported by TensorFlow. The distributed strategy API offers more flexibility by enabling machine learning training tasks in a distributed fashion without even changing the model definition.

- **Heterogenous in nature:** The TensorFlow Extended (TFX) pipeline is used for the full production of ML algorithms in server levels. TensorFlow Lite is used on mobile and edge devices for executing inference. TensorFlow.js is used to support training and execution in JavaScript environments. Overall, TensorFlow supports the training and deployment in any platform with any language.

- **Robust:** TensorFlow supports a wide range of libraries and models to form an ecosystem that is powerful enough to experiment complex algorithms, functional APIs, and model subclassing APIs. This in turn preserves the robust nature of TensorFlow with powerful models without compromising the performance.

- **Well documented:** The features, functions, classes, and modules of TensorFlow are well defined and documented in a more understandable manner to support developers in extending its features and to help users write applications effectively.

2.2.2 Modules of TensorFlow

The following are the modules of TensorFlow:

Module	Purpose
audio module	This module supports the encoding and decoding of tensors into WAV files and vice versa.
autodiff module	This module offers functions for automatic differentiation to measure the rate of change using derivatives of the mathematical functions used in machine learning.
autograph module	This module works with TensorFlow graphs, which can be viewed as computation nodes and edges as data flow diagrams (DFDs).
bitwise module	This module performs bitwise operations on tensor elements.

(continued)

Module	Purpose
compat module	This module helps in writing compatible code that works both in TensorFlow 1.x and 2.x and Python 2 and 3 using compatible functions.
config module	This module performs the configuration of logical or physical devices for experimental, optimizer, and threading purpose.
data module	The input data pipeline type specification and representation methods such as iterators, TFrecords, and MultilineText are offered by this module.
debugging module	This module helps to debug using functions such as `assert_equal`, `assert_less`, and `assert_greater` elementwise. It has functions to check or assert every attribute of tensor.
distribute module	This module helps in distributing the algorithm among multiple machines or devices.
dtypes module	This module works around the data types of tensors and converts/cast the tensors type.
errors module	This module handles the exception types and errors that are generated during execution.
estimator module	The estimator module does training, evaluation, prediction, and export. This module has all the experimenting and exporting methods.
experimental module	This module supports experimenting with deep learning packages, distributed tensors, and TensorFlow-TensorRTcompiler for inference on NVIDIA devices, etc.
feature_column module	This module works with the several types of columns such as numerical, categorical, embedding, hash, etc.

(continued)

Module	Purpose
graph_util module	This utility functions helps in manipulating the tensor graphs.
image module	This module has various methods to manipulate, encode/decode, and process the images.
io module	This module works with file I/O wrappers to support multiple file system implementations.
keras module	This module enables the access of the Keras API.
linalg module	This module has access to operations supported for linear algebra.
lite module	This module enables the access of TensorFlow Lite models and supporting methods.
lookup module	This module has functions and classes to initialize the lookup table, which contains key and value tensors with the corresponding index to access easily.
math module	This module allows access to various math functions such as arithmetic, trigonometric, complex number, reduction, scan, etc.
mlir module	This module works with multilevel intermediate representation.
nest module	This module works with nested data structures and operations on them.
nn module	This module supports access to primitive neural network operations such as pooling, padding, etc.
profiler module	This module has functions for creating, managing, and tracking the profiles. It offers tools for measuring the effective resource consumption models.

(continued)

Module	Purpose
quantization module	The module provides function for quantizing or reducing the precision of numbers, which may be the values of parameters. It also supports dequantize.
queue module	This module provides access to different queues and its operations such as FIFO queue, priority queue, random shuffle queue, etc.
ragged module	This module has methods to operate on ragged tensors, which are in nonuniform shapes.
random module	This module works with generation of random numbers and samples.
raw_ops module	This module provides direct access to all low-level TensorFlow operations.
saved_model module	This module has access to saved models in TensorFlow, which does not require original model building code to run.
sets module	This module has methods for TensorFlow set operations such as union, intersection, etc.
signal module	This module enables the signal processing operations such as discrete cosine transform, fast Fourier transform, etc.
sparse module	This module supports operations on sparse tensors and its representation.
strings module	This module especially works with string tensors.
summary module	This module supports operations for writing summary data that can be used for analysis.

(*continued*)

Module	Purpose
sysconfig module	This module gives us the information about the system configuration such as compilation and linker flags, C++ header files, and TensorFlow framework library directory location
test module	This module helps in understanding the testing of TensorFlow with its benchmarks.
tpu module	This module describes the options for tensor processing unit and accelerated linear algebra (XLA) compilation.
train module	This module works with training models and its associated methods.
types module	This module has the methods to convert union of all types into tensors.
version module	This module gives the version details of the compiler, `git`, `graph_def`, etc.
xla module	This module supports the XLA optimizing compiler to accelerate TensorFlow models.

2.3 PyTorch

PyTorch is a specialized machine learning and deep learning library that uses GPUs in addition to CPUs. It is a healthy competitor to Keras and TensorFlow and good at tensor manipulations, GPU acceleration, and highly optimized and automatic differentiation. Google released Keras in early 2015 to train neural networks with easy-to-use APIs. But the low-level features are unable to be customized. In late 2015, Google released TensorFlow, which gained popularity soon and serves as a back end for the Keras library. It also implemented more lower-level features for researchers and deep learning practitioners to undergo novel approaches in their research. But the implementations are not Python based and not able to be used spontaneously in implementations.

These drawbacks of Keras and TensorFlow were solved by the release of PyTorch. PyTorch implemented the low-level functions that are readily easy to use and customizable. Also, the functions are Python based, quite suitable, and easier to integrate. PyTorch represents the data in terms of arrays with the support of multidimensional representation called *tensors*. Release types of PyTorch packages (not only for PyTorch but for almost all packages) are stable, beta, and prototype. Stable releases are to be maintained for the long term, and no major performance issues are raised from the user community. Backward compatibility is also provided, and deprecated functions if any will be notified with every release. In beta releases, coverage will not be complete as the changes are performed based on user feedback, with no backward compatibility ensured. Prototype releases are exclusively for testing and feedback purposes.

2.3.1 Features of PyTorch

The following are the key features of PyTorch:

- **Easy to customize:** The libraries and subpackages in PyTorch are developed in a way to support customization in generating new optimizers, models, layer types, and architectures. PyTorch is easier for researchers and experts from various domains to implement applications and research algorithms.

- **Production ready:** TorchScript supports the transition between eager mode (suitable for coding and debugging but not advisable for production) and graph mode (supports real-time production). TorchServe serves as an accelerator in companion to TorchScript.

- **Distributed training:** PyTorch offers the distributed operation with a *distributed* subpackage that supports training and performance optimization in applications under production as well as research. The distributed environment not only provides performance improvement but also provides scalability.

- **Robust ecosystem:** A well-structured set of tools, frameworks, modules, and libraries in PyTorch leads to the formation of a robust ecosystem to develop the applications in computer vision, natural language processing, recommendation systems, etc.

- **Cloud support:** Major cloud platforms support PyTorch, and the applications can be implemented with the features of the cloud for development, vigorous testing, and deployment. It also enables easy scaling of resources and the dimension of applications in every perspective.

2.3.2 Libraries in PyTorch

The following are the PyTorch libraries:

Library	Usage
TorchAudio	It deals with signal and audio processing. Loading and storing waveform tensors from and to files, media stream encoding and decoding, synthesizing, filtering, etc., are the operations supported.
TorchData	The data pipelines (a series of processing steps on data) are one of the key features in data science and analytics. TorchData contains the data loading primitives for easy construction and usage of data pipelines.
TorchRec	It is a PyTorch library that deals with recommendation systems that are very large. General parallelism and sparsity primitives are the key features of the TorchRec library to train models with large datasets shared across many GPUs.
TorchServe	It offers easy and flexible tools to deploy PyTorch models in production. In addition, it provides easy versioning along with built-in or custom handlers, multiple models in one instance, etc.
TorchText	This library deals with sentiment analysis, sequence tagging, classification, corresponding data processing utilities, and datasets.
TorchVision	It deals with computer vision-based manipulations on images and videos such as transformation and augmentation, segmentation, object detection, classification, etc.
Pytorch on XLA Devices	This library supports methods and enables the PyTorch execution on Accelerated Linear Algebra (XLA) devices.

2.3.3 Modules in PyTorch

The following are the modules in PyTorch:

torch	This package defines the data structure for tensors of multidimensional and mathematical operations over tensors.
torch.nn	This package is a neural network library that contains containers, various layers such as convolution, pooling, padding, normalization, and activations and functions.
torch.Tensor	This package represents the tensor as a multidimensional matrix with elements of the same data type and with basic operations on tensors.
torch.amp	This package supports automatic mixed precision (amp). Methods such as autocast and GradScaler are mostly used.
torch.autograd	This package offers automatic differentiation of arbitrary valued functions.
torch.library	This package helps to create new libraries by extending core PyTorch's library.
torch.cuda	This package uses graphical processing units (GPUs) for processing and computation to support CUDA-type tensors.
torch.mps	The Metal Performance Shaders (MPS) back end is used to accelerate training in GPUs. This package provides an interface to the MPS back end.
torch.backends	The back ends supported in PyTorch are `cuda`, `cudnn`, `mps`, `mkl`, `mkldnn`, `openmp`, `opt_einsum`, and `xeon`. This package manages and controls the back ends.

(*continued*)

torch.distributed	This package offers three distributed training and communication models: Distributed Data-Parallel Training (DDP), RPC-Based Distributed Training (RPC), and Collective Communication (c10d) for implementing distributed applications using PyTorch.
torch.distributions	This package supports probability distributions and sampling. Stochastic computation graphs and stochastic gradient estimators can be built for optimizers.
torch.fft	This package deals with the discrete fourier transforms and the methods involved in its operation.
torch.func	This package supports composable function transforms that take input as a function in one form and produces an output function to give different quantities.
torch.futures	This package supports the Future type, which is used in the RPC framework. The Future type has a set of utility functions and enables asynchronous execution.
torch.fx	This package offers a toolkit that has three components to transform nn.Module instances. The three components are symbolic tracer, intermediate representation, and Python code generation, which makes Python-to-Python transformation.
torch.hub	This package supports a pretrained model repository that can be reproduced for research or application development purposes.
torch.jit	This package deals with the generation of the TorchScript program code from the Python code. The generated TorchScript code can be executed independently without a Python environment.

(continued)

torch.linalg	This package deals with linear algebra operations such as decompositions, solvers, inverses, matrix functions, etc. It also has methods to handle extremal values that may overflow the storage of the data type used.
torch.monitor	This package offers an interface to look after and record the events and counters as logs.
torch.signal	This package deals with signal windows that can compute Barlett, Blackman, Gaussian, etc., windows for processing the subset of datasets.
torch.special	It computes the value of special functions by taking input as tensors and outputting them as tensors. The special functions are Bessel, gamma, entropy, error, complementary error, logistic sigmoid, inverse error, etc.
torch.overrides	This package helps to create custom types with functions that match the tensor type to emulate tensor type. It redirects the call to a custom implementation whenever any Torch functions are called.
torch.package	It offers the methods to create packages containing PyTorch code that can be executed, saved, and shared on different platforms or can be deployed.
torch.profiler	It offers a collection of performance metrics used to understand the model, their input shape, stack trace, execution trace, etc.
torch.onnx	Open Neural Network eXchange (ONNX) is an open standard for representing ML and DL models. This package helps to export PyTorch models to the ONNX format.

(continued)

torch.optim	This package offers functions that implement well-known optimizers such as Adadelta, Adagrad, Adam, RMSprop, etc. It also has interfaces to integrate new optimizers in the future.
torch.random	This package supports the generation of random numbers by forking the random number generators (RNGs). It also has operations supporting the random number generation such as setting the seed for generation, setting the RNG state, etc.
torch.masked	This offers methods for masking or including the subset of input. Masked tensor is a subclass of the `Tensor` class, which requires a mask as one of the inputs.
torch.nested	This package supports the packing of more than one tensor into a single data structure with the constraint that all the input tensors is of the same dimension.
torch.sparse	This package deals with sparse tensors and the generation of a sparse tensor from the dense tensor. The generated sparse tensor has more zero-valued elements.
torch.Storage	This package deals with the storage class with each data type. It also supports methods to move storage to shared memory, pinned memory, etc.
torch.testing	This offers methods for testing and assertions to check the actual and expected values. It may raise errors such as ValueError, AssertionError, TypeError, etc.
torch.utils	This package provides common utilities such as setting and getting the values of model parameters. Also, many more subpackages are implemented under `torch.utils`.

2.4 SciPy

SciPy is a Python package developed for scientific computation–based applications. It offers statistical, signal processing, and optimization-based functions. It was developed by Travis Olliphant who developed NumPy. So, SciPy uses NumPy as its foundation. The functions that are offered in SciPy are additional support and optimized versions of frequently accessed functions in NumPy. The SciPy collection of mathematical algorithms and functions are widely used in computation-intensive applications.

SciPy also offers various functions and classes for visualizing data to support decision systems and real-time analysis. It has rich user-friendly numerical functions for numerical computations, integration, differentiation, and optimization. SciPy is highly associated with NumPy, and the arguments and return types of SciPy functions are mostly NumPy arrays.

2.4.1 Features of SciPy

The key features of SciPy are as follows:

- **Easy and fast:** SciPy programs are easy to code and execute significantly faster than any other alternatives as they are developed with an extension of C that can manipulate large datasets in a very short time.

- **Built-in functionality:** SciPy has a huge extent of built-in functions. It does not require any additional libraries for image processing or simple matrix manipulation.

- **Parallel programming:** SciPy classes, methods and subpackages, and modules are developed in a manner to support parallel programming even though it has no direct function primitives for achieving parallelism.

- **Better approach for optimization:** SciPy contains various optimization methods that solve difficult optimizations in an easier manner.

- **Multiple-platform support:** SciPy supports execution on numerous platforms such as Windows, Linux, macOS, etc. A range of devices and operating systems are supported with its cross-platform development functionality.

2.4.2 Modules in SciPy

The following are the modules:

Modules	Purpose
scipy.cluster	The clustering algorithms are used in many applications such as compression, information theory, etc. This module supports clustering algorithms such as K-means, vector quantization, hierarchical, etc.
scipy.constants	This module provides almost all the mathematics and physics constants such as Pi, golden ratio, speed of light, electric constant, etc.
scipy.fftpack	The Fourier transform is used to study the behavior of time domain signals in the frequency domain. This module contains methods to compute Fourier transforms and its variants.
scipy.integrate	This module offers methods for computing various integrations such as general integration, general multiple integration, gaussian quadrature, Romberg integration, fast integration, ordinary differential equations, etc.

(continued)

Modules	Purpose
scipy.interpolate	Interpolation creating a new data point from a known set of discrete data points. This module offers methods to compute interpolations such as univariate, multivariate, 1D splines, 2D splines, etc.
scipy.io	Scipy.io provides a rich set of classes and methods to read and write data in various file formats such as matlab files, matrix market files, netcdf files, idl files, unformatted Fortran files, Harwell-Boeing files, wav sound files, and arff files.
scipy.linalg	This module supports the linear algebra operations such as eigenvalue problems, decompositions, matrix functions, matrix equation solvers, sketches and random projections, special matrices, low-level routines, etc.
scipy.ndimage	This module offers image processing functionality not only for 2D arrays but also for the higher dimensions such as medical and biological imaging. It includes functions such as linear and nonlinear filtering, binary morphology, B-spline interpolation, and object measurements.
scipy.odr	This module mainly resolves the measurement errors that exist in independent (explanatory) variables and not only in dependent variables. Orthogonal Distance Regression (ODR) offers Ordinary Least Squares (OLS), which defines fitting procedures to treat the data for explanatory variables as fixed, i.e., not subject to error of any kind.
scipy.optimize	This module provides functions for minimization/maximization subject to constraints. It has solvers such as scalar function optimization, local and global optimization, root finding, least squares and curve fitting, linear programming, assignment problems, utilities, etc.

(continued)

Modules	Purpose
scipy.signal	This module offers signal processing functions such as convolution, B-splines, filtering, filter designing, continuous time and discrete time linear systems,waveforms, wavelets, window functions, peak finding, spectral analysis, etc.
scipy.sparse	This module covers the sparse matrix earlier and is now switched to an array interface compatible with numpy arrays. The seven sparse matrix types are csc_matrix: Compressed Sparse Column format, csr_matrix: Compressed Sparse Row format, bsr_matrix: Block Sparse Row format, lil_matrix: List of Lists format, dok_matrix: Dictionary of Keys format, coo_matrix: COOrdinate format (aka IJV, triplet format), dia_matrix: DIAgonal format.
scipy.special	This module defines special functions such as airy functions, elliptic functions, various bessel functions, struve functions, information theory functions, raw statistical functions, gamma functions, etc.
scipy.stats	This module offers functions for probability distributions such as continuous, discrete, multivariate, etc. It also offers summary statistics, frequency statistics, hypothesis tests and related functions, etc.

2.5 Theano

Theano is a Python library enabling computationally intensive applications and research on a large scale. The multidimensional mathematical arrays and expressions are efficiently defined, manipulated, optimized, and evaluated with the Theano Python library functions.

The characteristics of computer-based algebraic system and the characteristics of optimized compiler are combined in Theano functions to evaluate critical mathematical operations.

The amount of computation, analysis, and compilation overhead involved can be minimized while manipulating the complex mathematical symbolic features such as automatic differentiation. Also, Theano's compiler offers various optimizations with a gamut of complex symbolic expressions such as constant folding, arithmetic simplification, memory aliasing to avoid repeated calculation, loop fusion for elementwise subexpressions, merging of subgraphs to avoid redundant computation, etc.

2.5.1 Features of Theano

The key features of Theano are as follows:

- **Dynamic C code generation:** Theano can generate customized C code that is executed in GPUs that attain speeds by many orders of magnitude greater than the manually coded C implementations executing on CPUs.

- **Effective symbolic differentiation:** Theano has methods to generate and compute derivatives for mathematical functions of one or many inputs. It also supports automatic differentiation, which speeds up the execution.

- **Speed and stability optimizations:** Many mathematical expressions are complex, and generating values for such expressions when larger variables are involved is time consuming and error prone. This feature is used to avoid such errors and exceptions when computing expressions such as $\log(1 + \exp(x))$ for larger x.

- **Tight integration with NumPy:** Many modules internally use NumPy and ensure the support of NumPy arrays and functions from/to external interfaces. The numpy.ndarrays type is used in the internal operations of Theano-compiled functions.

- **Transparent use of GPU:** CPUs generally support the floating-point operations of lesser word size (such as float32). Higher bit size data types and operations are defined and supported in Theano. GPUs are involved in those kinds of computations.

- **Wide-ranging unit-testing and self-verification:** Theano has a rich set of methods and tools to detect and analyze the bugs or errors generated during execution. Various unit testing error detection modules exist in Theano.

2.5.2 Modules in Theano

The following are the modules:

Module	Purpose
compile	This module deals with the transformation of expression graphs to functions, which can handle and manage shared variables, common operations, and optimizations on functions; interfaces for graphs into callable objects; and various modes for compilation.

(continued)

47

Module	Purpose
config	This module governs the configuration of Theano and hence controls the behavior of Theano. The value for attributes are set by default using `Theano.config.<Property>`. This in turn overridden by the `THEANO_FLAGS` environment variable. The next level of modifying the attribute value is through `.theanorc` file.
d3viz	This module offers the interactive visualization of computation graphs. It creates an HTML file instead of static pictures to open in any web browsers to view. It supports zooming, positioning nodes manually or automatically, editing node labels and retrieving node and edge information, and exploring nested graphs.
gof	This module provides an interface for accessing the graphs, its containers, types, etc. It also provides support for operating on graphs using toolbox and utility functions.
gradient	This module acts as a driver for gradient calculations. Symbolic gradient are computed with the functions `gradient.grad()` and `grad_sources_inputs()`. It provides all gradient-related functions.
misc.pkl_utils	This module provides various operations and tools for managing serialization. It pickles (treats, preserves, and cleans) the object pointing to the zipped file after its access, with the condition that the zip folder should contain at least one file compatible with Numpy's functions.
printing	This module enables and supports the printing of intermediate values in computation that cannot be printed using normal print statement in Python. The order of printing depends on the graph.

(continued)

Module	Purpose
sandbox	This module offers predefined experimental codes and functions for CUDA backend, GPU array backend, linear algebra ops, neighbors in convolutional nets, and Multiple Recursive Generator (MRG) random number generator.
scalar	This module governs the operations and expressions related to special data types such as iscalar, fscalar, etc.
scan	This module provides the functionality of manipulating array elements through loops such as a simple loop with accumulation, iterating over the first dimension of a tensor, simple accumulation into a scalar, etc.
sparse	This module is the same as the tensor module, which governs sparse matrices, but sparse matrices do not store data in a contiguous array. It supports two compressed sparse formats, csc and csr, respectively, based on columns and rows. It can be manipulated with attributes such as data, indices, indptr and shape, etc.
tensor	This module comprises tensor manipulation and processing functions. There are various types of symbolic expressions for tensors such as nnet, raw_random, signal, etc.
typed_list	This data type is added in release 0.7 of Theano. This module creates and manipulates lists in which all elements are of the same data type.

2.6 Pandas

Pandas is a data processing–based Python package that provides efficient data structures called *series* and *dataframe*. A series handles one-dimensional data, and a dataframe deals with higher dimensions. A dataframe mainly works with relational or table data that constitutes heterogenous typed columns. It forms a basic but powerful building block of data for performing real-world data analysis in Python.

Data in any form such as ordered/unordered time series, arbitrary rows and columns type, statistical records, etc., can be easily represented in series or dataframe. Like many other packages, Pandas is also built on top of the NumPy to contribute to scientific applications. It supports robust data input and output tools that can be loaded from or to any kind of source such as Excel files, flat files, databases, and sophisticated file systems.

2.6.1 Features of Pandas

The key features of Pandas package are as follows:

- **Prudent handling of missing data and Size mutability:** Pandas handles missing data easily by representing it as NaN (Not a Number). Though it is of type floating point, it is also used for non-floating-point data. Also, columns can be deleted or inserted from or to the dataframe and higher-dimensional objects.

- **Automatic data alignment:** The data in the dataframe will be automatically aligned during computations. It can also be explicitly aligned with a set of labels.

- **Excellent data processing modules:** Pandas offers functions to perform split-apply-combine operations on datasets, for both aggregating and transforming data. It enables conversion of ragged, differently indexed data in other Python and NumPy data structures into dataframe objects. It also supports intelligent label-based slicing, fancy indexing, subsetting of large datasets, intuitive merging and joining datasets, flexible reshaping and pivoting of datasets, and hierarchical labeling of axes.

- **Effective support for various file formats:** The various file formats for loading the data to the dataframes are from Excel files, comma separated values (CSV) or delimited files, and data from databases. Pandas also supports the loading of data from ultrafast Hierarchical Data Format Version 5 (HDF5).

- **Time-series functionality:** The periodic generation of data and the time-series data functions operated on data are supported. The moving window statistics, shifting of data, and handling of lagging data are supported in Pandas.

2.6.2 Modules in Pandas

The following are the modules:

Modules	Purpose
pandas.errors	This submodule handles the warnings and exceptions raised in the applications by Pandas functions.
pandas.plotting	This module provides the interface for plotting of various graphs in Pandas.
pandas.testing	This module offers the functions that are used for testing when Pandas objects are involved.
pandas.api.extensions	This module supports the extensions in the form of functions or classes.
pandas.api.indexers	This module provides functions and classes for accessing objects through indexers.
pandas.api.interchange	This module governs the dataframe interchange protocol. This enables the conversion from Pandas and cuDF to the Vaex, Koalas, Modin, and Ibis dataframes.
pandas.api.types	This module has the datatypes and its corresponding functions, which are supported in Pandas.
pandas.io	This module supports the reading and writing of data from or to Flatfile, Clipboard, Pickling, Excel, HTML, XML, JSON, Latex, HDF5, Feather, etc.
pandas.tseries	This module offers integration of time series and data through functions and classes.

2.7 Matplotlib

Matplotlib is one of the fundamental Python packages to visualize the data through static or interactive charts. It was developed by John D. Hunter. Matplotlib has two application interfaces (APIs): an explicit "Axes" and an implicit "pyplot." The Axes interface works on `Figure` objects and constructs the visualization in stages. It is also called an object-oriented interface as it uses methods of the `Figure` object to create an `Axes` object, which in turn enables the call to plot or drawing methods. The Implicit pyplot API creates the `Figure` and `Axes` objects by itself and enables the user to call the plot method directly. Though the `plot()` method is available in the data handling libraries such as Pandas, xarray, etc., Matplotlib offers functions specific for the visual demonstration of data and improves the understanding of data.

Matplotlib initially manages 2D plots, and the functions and utilities are defined based on two dimensions in its 1.0 release. Successively, three-dimensional plotting utilities were also implemented on top of the 2D display. It enables a comfortable way of visualizing three-dimensional plots with the set of tools defined under the mplot3d toolkit. Hence, the 3D plots are generated in applications by importing matplot3d with `from mpl_toolkits import mplot3d`.

2.7.1 Features of Matplotlib

The following are some of the key features of Matplotlib:

- **Open source and simple to visualize large data:** Matplotlib supports various types of data representation such as pie charts, bar charts, histograms, graphs, scattered plots, line plots, etc. It provides a simple and elegant approach for accessing large amounts of data.

- **Flexible, extensive, and customizable:** Matplotlib offers flexibility in adapting any kind of data and generates its visualization. Also, with its gamut of plots support, it can be customized to the required form.

- **Powerful and easy to navigate:** The data visualization feature of matplotlib can be utilized in many Python scripts, shells, Jupyter notebooks, and web application servers. It is not necessary to be a software professional to understand the features of matplotlib as it is easier to produce professional visualizations.

- **Generates high-quality plots and advanced visuals:** The main focus of Matplotlib is generating good plots and better visuals, it produces high-quality images that have various plots and visuals in the form of PDF, PGF, PNG, etc.

- **Executable on different platforms:** As Matplotlib supports platform independence, it is able to run on different platforms such as macOS, Windows, or Linux and produce high-quality results.

2.7.2 Modules in Matplotlib

The following are the modules:

Module	Purpose
matplotlib.afm	This module is used as an interface for Adobe font metrics files. But its use is deprecated.
matplotlib.animation	Animation in Matplotlib can be achieved with this module. It contains two main submodules. The writer class generates animations, and the helper class helps in the writer registry to map writer and class passes string to `animations.save`.
matplotlib.artist	This module serves as a base class for all visible elements in the figure.
matplotlib.axes	This class governs plotted data, labels, title, legend, etc. Its methods are used to interface and manipulate the plots.
matplotlib.axis	This module manages the x-axis and y-axis of the plots and its ticks.
matplotlib.backend_ bases	This module serves as the base for implementing graphics context that interface with the matplotlib backend.
matplotlib.backend_ managers	This module act as a manager for the user actions that trigger the functions to execute.
matplotlib.backend_ tools	This module defines the primitives for tools such as stateless tool, toggle tool, copy to clipboard tool, etc.
matplotlib.backends	This module governs all the backends such as the Python shell, Jupyter notebooks, graphical user interface-based applications, web application, etc., through which matplotlib provides its visualization.

(continued)

Module	Purpose
matplotlib.bezier	This module provides utility for working with Bezier paths.
matplotlib.category	This module generates graphs with any axis as categorical variables/strings.
matplotlib.cbook	This module provides various utility functions and classes.
matplotlib.cm	This module supports built-in color maps.
matplotlib.collections	This module holds the classes for generating plots for the large collections of objects.
matplotlib.colorbar	This module enhances the visualization by mapping the scalar values to colors.
matplotlib.colors	This module has the functionality of converting a string or numeric representation of color into RGB or RGBA sequences.
matplotlib.container	This module serves as a container for the collection of semantically related items in the plots, for example, set of bars in bar plot.
matplotlib.contour	This module generates contour plots that illustrates 3D surface in a 2D format with constant Z slices.
matplotlib.dates	This module builds plots by mapping the dates on the axis.
matplotlib.docstring	This module is used to describe the components in the plots using strings over it. But this module is deprecated.
matplotlib.dviread	This module helps in reading the DVI files output by LaTex or Tex.
matplotlib.figure	This module consists of `Figure`, `SubFigure`, and `SubplotParams` that hold all plot elements.
matplotlib.font_manager	This module governs the fonts by finding, using, and managing across the platforms.

(*continued*)

Module	Purpose
matplotlib.ft2font	This module provides classes and functions to use and support ft2 TrueType fonts.
matplotlib.gridspec	This module provides the grid view in the figure with multiple axes.
matplotlib.hatch	This module supports the creation of hatch patterns with its various classes and functions.
matplotlib.image	This module supports image handling, reading, scaling, various image display options, etc.
matplotlib.layout_engine	This module handles all the arrangements and organization of components of the plots to provide a pleasing layout.
matplotlib.legend	This module offers classes and functions to draw a legend of the axes.
matplotlib.legend_handler	This module interfaces the low-level API as callable object to manage various legend types.
matplotlib.lines	This module supports classes and functions that generates 2D lines with various style, size, color, and markers.
matplotlib.markers	This module governs the various markers used by plots. The marker types are point, pixel, circle, triangle_down, triangle_up, etc.
matplotlib.mathtext	This module parses a math syntax and renders it to the matpotlib back end.
matplotlib.mlab	This module has functions with the same name in Matlab commands for numerical operations.
matplotlib.offsetbox	This module defines the positional relation between parent and child components and the package among them with offsets.

(continued)

Module	Purpose
matplotlib.patches	This module supports patches in any object of the figure with face color and edge color.
matplotlib.path	This module enables the drawing of ploy lines and serves as base class for all vector drawings.
matplotlib.patheffects	This module provides functions to draw multistage to any artist object in the visualization.
matplotlib.pyplot	This module is for generating interactive plots and programmatic plot generations.
matplotlib.projections	This module governs the mapping or transformation of data coordinates to display coordinates.
matplotlib.quiver	This module supports the ploting of vector fields. Two plots, namely, `quiver` and `barb`, plots are currently under it.
matplotlib.rcsetup	This module offers the functions and classes for runtime configuration (RC) settings for customization of Matplotlib plots.
matplotlib.sankey	This module enables the visualization of Sankey diagrams.
matplotlib.scale	This module defines the data values on the axis. That can like logarithmic, linear, etc.
matplotlib.sphinxext	This module defines the functions and modules to get input as `.rst` (ReStructure Text) files and convert into HTML. It supports sphinx, which is a tool to automate documentation in an elegant manner.
matplotlib.spines	This module manages the spines, which are lines connecting the boundaries of data area with the tick marks on axis.

(continued)

Module	Purpose
matplotlib.style	This module works with various styles that dictates the visual appearance of the plots. The styles can be specified using RC parameters.
matplotlib.table	This module is used to generate tables for storing values in a grid of cells from texts.
matplotlib.testing	This module is developed for testing the figures and images by comparing them.
matplotlib.text	This module generates the text in the figure. It has x, y, and string text as its base parameters.
matplotlib.texmanager	This module supports TeX expressions such as LaTeX, PS, PDF backends, etc., in figures.
matplotlib.ticker	This module determines the location and appearance of the major and minor ticks along the axes.
matplotlib.tight_bbox	This module governs the functions and classes to generate bounding boxes as axis aligned rectangles denoted by a tuple of four integers and it is deprecated.
matplotlib.tight_layout	This module enables the layout with not only plots but also subplots with convenient arrangement of them.
matplotlib.transforms	This module depicts the geometric transformations to decide the final position of all component objects in the figure. Transforms are denoted by a tree in which every object depends on its children in its geometric value.

(continued)

Module	Purpose
matplotlib.tri	This module generates unstructured triangles automatically generated using a Delaunay triangulation.
matplotlib.type1font	This module supports Type 1 font. It also supports slant font and extend font transformation. This module is deprecated.
matplotlib.typing	This module instructs color type, RGB color type, RGBA color type, line style type, fill style type, draw style type, etc.
matplotlib.units	This module supports functions and classes for units and its conversions.
matplotlib.widgets	This module supports the working of GUI backends.
matplotlib._api	This module is for developers to create helper functions for managing API.
matplotlib._enums	This module defines the sets of values for many attributes and parameters of figures in Matplotlib.
mpl_toolkits.mplot3d	This module adds 3D plotting feature to the Matplotlib such as scatter, surface, line, mesh, etc.
mpl_toolkits.axes_grid1	This module governs the positions of multiple fixed aspect axes to display plots and figures through helper classes.
mpl_toolkits.axisartist	This module has the feature of generating curvilinear grids. Axis lines, ticks, and labels are different from normal Matplotlib axis class.

2.8 Scikit-learn

Scikit-learn is a machine learning library that has implemented supervised and unsupervised learning. Besides, it is an open-source library distributed under the three-clause BSD (Berkley Source Distribution) license and supports various tools such as model fitting and predicting, model selection, model validation, transforming and preprocessing, pipelines, and many more utility functions. The powerful automatic parameter search is provided by Scikit-learn to decide the best parameter combinations.

2.8.1 Features of Scikit-learn

The following are some of the key features of Scikit-learn:

- **Open source:** The Scikit-learn package is open source for noncommercial purposes and available under the BSD license for commercial purpose.

- **Built-in dataset:** There are handful of built-in dataset available in Scikit-learn such as toy datasets (iris, diabetes, digits, physical exercise Linnerud, wine, breast cancer), real-world datasets (Olivetti faces from AT & T labs, 20 newsgroups, Labelled Faces in the Wild (LFW) people, California housing), and generated datasets (artificial datasets created with various random sample generators). These datasets are suitable for beginners.

- **Efficient ensemble methods:** Scikit-learn supports the combination of various supervised models and blends the benefits in the predictions to get better results. It supports bagging, boosting, random forest, etc.

61

- **Precise cross validation:** Usually data in classification, regression, etc., is divided into training and test data based on a ratio such as 80:20, 70:30, etc. Instead, the cross validation is used to uncover the accuracy of supervised learning algorithms precisely on unseen data.

- **Plenty of feature extraction and selection methods:** Scikit-learn supports feature extraction such as patch extraction, connectivity graph in images, bag of words representation, sparsity, term frequency(tf) – inverse document frequency (idf) weighting, feature hashing, and feature selection methods such as univariate selection, recursive elimination, L1 based, tree based, pipeline based, etc.

2.8.2 Modules in Scikit-learn

The following are the modules:

Module	Purpose
sklearn.base	This module provides base classes and utility functions for all estimators.
sklearn.calibration	This module helps in the calibration of predicted probabilities.
sklearn.cluster	This module supports unsupervised clustering algorithms.
sklearn.compose	This module consists of composite estimators or meta estimators for generating composite models with transformers.
sklearn. covariance	This module includes the methods and functions to estimate the covariance of the features defined in data points.

(continued)

Module	Purpose
sklearn.cross_ decomposition	This module incorporates the algorithms for cross decomposition of supervised estimators for the partial least square family such as regression, dimensionality reduction, etc.
sklearn.datasets	This module has functions to load the datasets and fetch popular datasets and also supports artificial intelligent data generators.
sklearn. decomposition	This module encompasses matrix decomposition algorithms such as principal components analysis (PCA), non-negative matrix factorization (NMF), independent component analysis (ICA), etc.
sklearn. discriminant_ analysis	This module focuses on linear and quadratic discriminant analysis.
sklearn.dummy	This module consists of fake estimators to compare the model estimators based on parameters such as mean, median, quantile, and constant.
sklearn.ensemble	This module contains methods to combine the predictions of more than one base estimators of a given algorithm over single one. Examples are random forests and gradient boosted trees.
sklearn. exceptions	This module covers all the errors classes and warnings defined across the Scikit-learn.
sklearn. experimental	This module enables the methods and classes defined to use the experimental features of estimators. But care should be taken in using them as it is not in a deprecation cycle.
sklearn.feature_ extraction	This module is used to extract features from raw data. It has methods to extract features from text and images.

(continued)

Module	Purpose
sklearn.feature_ selection	This module has methods to select specific features and eliminate other parts of the images using univariate filter selection methods and the recursive feature elimination algorithm
sklearn.gaussian_ process	This module supports the Gaussian process based algorithms such as regression, classification, etc.
sklearn.impute	This module focuses on missing values and its replacement with values, which should retain the information present in the dataset.
sklearn.inspection	This module inspects the performance of the models with various methods such as permutation of feature columns, partial dependence of a feature, decision boundary visualization, etc.
sklearn.isotonic	This module works with predictions of isotonic regression that fits for a nondecreasing real function to 1D data, which is piecewise linear.
sklearn.kernel_ approximation	This module focuses on approximation of features mapping to existing kernels. The approximation used are Nystroem method, skewed Chi squared, radial basis function, etc.
sklearn.kernel_ ridge	This module has methods to implement kernel ridge regression that combines the ridge regression and classification with kernel trick. Hence, it learns linear and nonlinear function with the corresponding kernel.
sklearn.linear_ model	This module includes various linear models' implementation such as Lasso, Multitask Lasso, Elastic-Net, Orthogonal Matching Pursuit (OMP), logistic regression, Ordinary Least Squares etc.

(continued)

Module	Purpose
sklearn.manifold	This module covers the manifold learning, which performs dimensionality reduction for nonlinear data.
sklearn.metrics	This module concentrates on metric for determining the quality of predictions. The metrics are classification metrics, multilabel ranking metrics, regression metrics, clustering metrics, etc.
sklearn.mixture	This module allows learning of Gaussian mixture models such as diagonal, spherical, tied and full covariance matrices.
sklearn.model_ selection	This module helps in choosing a model based on cross-validation, learning curve sections, tuning of hyper parameters, etc.
sklearn.multiclass	This module implements the multiclass classification strategies such as one versus all, one versus rest, one versus one, and output codes for error correction.
sklearn. multioutput	This module provides meta-estimators, which receives a base estimator as input and produces multioutput estimators from single output estimators.
sklearn.naive_ bayes	This module consists of naïve Bayes algorithms based on Bayes theorem with naïve feature independence assumptions.
sklearn.neighbors	This module has the functionality to support neighbor-based algorithms in both supervised and unsupervised algorithms.
sklearn.neural_ network	This module supports the neural network models of both supervised and unsupervised such as Bernoulli restricted Boltzmann machine (RBM), multilayer perceptron regressor, multilayer perceptron classifier, etc.

(continued)

Module	Purpose
sklearn.pipeline	This module has utilities and functions to implement the combination of estimators as a composite model, which consists of chain of transforms and estimators as a pipeline.
sklearn.preprocessing	This module focuses on the preprocessing of data such as standardization, normalization, encoding, nonlinear transform, discretization, imputation, etc.
sklearn.random_projection	This module implements random projection, which is an efficient method to reduce the dimension trading with the amount of accuracy for faster execution and smaller models.
sklearn.semi_supervised	This module provides algorithms for semi-supervised learning with less labeled data and large amount of unlabeled data. It also supports label propagation and label spreading.
sklearn.svm	This module has implemented various support vector machine (SVM) algorithms for regression, classification, and outlier detection.
sklearn.tree	This module focuses on decision tree models and extremely randomized tree models for classification and regression.
sklearn.utils	This module provides all kinds of utility functions that are required for implementing and validating the algorithms of Scikit-learn.

2.9 Seaborn

Seaborn is a library for generating statistical graphics, which supports the switching between different visual representation to provide better understanding of data. Its development is based on the Matplotlib and Pandas data structures and functions. Its plotting methods operate either on the the Pandas dataframe or on arrays that hold the entire dataset and accomplish the meaningful mapping and statistical aggregation to generate more elaborative plots. This package supports datasets of different formats and its corresponding possibility of different types of plots.

2.9.1 Features of Seaborn

The key features of seaborn package are as follows:

- **Built-in themes:** The visualization of plots are enhanced automatically with built-in themes in the Seaborn package. The beginners who are not very aware of styles and themes can make use of this feature.

- **Visualizing univariate and bivariate data:** The data of any type can be visualized independently or in relation with other attributes in the dataset. It supports univariate, bivariate, and their relationships.

- **Fitting in and visualizing various models:** The learning models are generated and visualized for how well they fit similar datasets. Various parameters and plots are supported to visualize the model fitting.

- **Statistical time-series data plots support:** The time-series data is continuously generated, and it can be plotted as line plots, density plots, heat maps, scatter plots, etc. It is highly effective in understanding temporal relationship with system generating the time series data.

2.9.2 Modules in Seaborn

The following are the modules:

Module	Purpose
seaborn.objects	This module is released with the version 0.12. This is an interface for creating Seaborn plots. It provides a flexible API for transforming the data into plots.
seaborn.relplot, seaborn.scatterplot, seaborn. lineplot, seaborn.displot, seaborn.histplot., seaborn. kdeplot, seaborn.ecdfplot, seaborn.rugplot, seaborn. distplot, seaborn.catplot, seaborn.stripplot, seaborn. swarmplot, seaborn.boxplot, seaborn.violinplot, seaborn.boxenplot, seaborn.pointplot, seaborn. barplot, seaborn.countplot, seaborn.lmplot, seaborn. regplot, seaborn.residplot, seaborn.pairplot, seaborn. jointplot	These modules generate various plots and are broadly categorized under three types: relationship between variables, distribution of variable values, and relationship of categorical values.
seaborn.heatmap seaborn.clustermap	These modules depict the representation of data in terms of colors to show the range of values and clustered over a given space.
seaborn.FacetGrid, seaborn.PairGrid, seaborn. JointGrid	These modules express the conditional relationship, pairwise relationship, and bivariate with marginal univariate plots.

(*continued*)

Module	Purpose
seaborn.set_theme, seaborn.set_style, seaborn. set_context, seaborn.set_color_codes, seaborn. reset_defaults, seaborn.reset_orig, seaborn.set, seaborn.set_palette	These modules govern the style and appearance of the plots. The colors of all components can be set using these methods and classes.
seaborn.axes_style, seaborn.plotting_context	These modules are used to get the values of the parameters to control the plots and axes style.
seaborn.color_palette, seaborn.husl_palette seaborn.hls_palette, seaborn.cubehelix_palette seaborn.dark_palette, seaborn.light_palette seaborn.diverging_palette, seaborn.blend_palette seaborn.xkcd_palette, seaborn.crayon_palette seaborn.mpl_palette	These modules are providing the functions and classes that are returning the list representing various formats of color palette such as RGB, Hue Saturation Light, etc.
seaborn.choose_colorbrewer_palette seaborn.choose_cubehelix_palette seaborn.choose_light_palette seaborn.choose_dark_palette seaborn.choose_diverging_palette	These modules are helping to choose the color from the given type of palette through various methods and classes.
seaborn.despine	This module is used to remove the top and right spines from plots. Spines are lines that connect the axis ticks and boundaries of the data area.

(continued)

Module	Purpose
seaborn.move_legend	This module is used to re-create the legend in a new location.
seaborn.saturate	This module is used to retrieve the fully saturated color of any plot or component with the same hue.
seaborn.desaturate	This module is responsible for reducing the saturation channel of a color with a predefined percent.
seaborn.set_hls_values	This module is used to manipulate the hue, saturation, and light channels of a color.
seaborn.load_dataset	This module enables the loading of a sample dataset from the Internet. Hence, this module requires the Internet.
seaborn.get_dataset_names	This module reports the existing list of datasets. This module also requires the Internet.
seaborn.get_data_home	This module retrieves the path of the example datasets available, and the path is used by load_dataset.

2.10 OpenCV

OpenCV is an open-source library that consists of a variety of methods for computer vision algorithms. It defines the basic data structure that includes multidimensional arrays, matrices, etc. It comprises libraries that implement image processing operations, video analysis, 2D features framework, calibration of camera and 3D reconstruction, object detection, high-level GUI, video I/O, etc.

Image processing operations consist of filtering, image transformation, color space conversions, etc. Video analysis consists of motion estimation, background filtration, object tracking, etc. A 2D feature framework comprises feature detectors, descriptors and descriptor matchers, camera calibration, and 3D reconstruction includes object pose estimation, stereo camera calibration, stereo correspondence algorithms, etc. Object detection has methods to detect the objects and predefined classes instances such as person, car, face, computer, pen, etc. A high-level GUI is an interface to access UI capabilities and its corresponding methods. Video I/O deals with the capturing of videos and coding and decoding of them.

2.10.1 Features of OpenCV

The following are some of the key features of OpenCV:

- **Open-source library:** The OpenCV library is open to use by the public for their applications. The code can be customized to meet the requirements of the business applications. Also, additional classes and modules can be integrated with it to provide extra functionality.

- **Faster execution:** The library is developed in C or C++. Python is an interpreter-based language, and hence it is slower in comparison with C/C++. But OpenCV enables the integration of C or C++ code by developing the wrapper classes for them to gain the benefit of faster execution.

- **Easily integrated with other languages:** OpenCV provides nearly 2,500 algorithms for manipulating images and videos. It offers methods to convert the OpenCV arrays into NumPy arrays to integrate with other Python libraries such as Matplotlib and SciPy. It also supports an interface for various languages such as C, Java, Python, etc., that makes integration easier.

- **Powerful and simple to code:** The methods in OpenCV makes the coding much simpler by providing gamut of functions for simpler to complex computations and algorithms. It also offers a powerful execution environment for highly computation intensive and scientific applications.

- **Rapid prototyping:** OpenCV applications can be prototyped quickly by integrating them with web frameworks such as Django, which supports rapid development of web-based applications.

2.10.2 Modules in OpenCV

The following are the modul.es:

Module	Purpose
core	This module provides the core functionality for implementing computer vision-based applications.
imgproc	This module works with all image processing methods such as filtering, transformations, segmentation, feature and object detection, etc.
imgcodecs	This module has methods to read and write images and supports various coding and decoding of images.
videoio	This module governs the video input and output by providing two main classes: video capture and video writer.
highgui	This module supports the instant interface for creating and manipulating image windows as OpenCV is meant for providing a range of user interfaces (UIs) from base to full rich UIs.
video	This module is tailored to track objects in video and do analysis on motion videos.
calib3d	This module holds the methods involved in calibrating the camera to obtain 3D scene by projecting a 3D point into the image plane using pinhole camera model and perspective transformation.
features2d	This module works with object labeling, descriptor matchers, feature detection, and description.

(*continued*)

Module	Purpose
objdetect	This module is used to detect objects with a wide variety of detection methods such as deep neural network (DNN) face recognition, barcode detection, QR code detection, cascade classifier for object detection, etc.
dnn	This module builds a deep neural network module that contains an API for layer creation, modifying comprehensive neural networks from layers, etc. It also has a set of built-in layers.
ml	This module consists of functions and classes for machine learning algorithms such as regression, classification, clustering, etc.
flann	This module is optimized for Fast Library for Approximate Nearest Neighbors (FLANN). It supports nearest neighbor search for higher dimensional features in large datasets.
photo	This module is used for photo processing algorithms such as denoising, contrast preserving decolorization, high dynamic range (HDR) imaging, inpainting, seamless cloning, etc.
stitching	This module provides stitching pipeline with all building blocks and processing steps. This can be customized to the application requirement by removing and enhancing the processing steps.
gapi	This module is Graph-API (GAPI) for processing regular images in a faster and more portable manner.

2.11 Summary

This chapter provided an overview of the Python packages that are widely used to implement learning algorithms. The packages are highly efficient and powerful in generating the models. The efficiency of the models is measured using the performance metrics supported by the packages. In the next chapter, we will discuss the supervised learning algorithms and their applications.

CHAPTER 3

Supervised Algorithms

3.1 Introduction

Supervised algorithms are grouped under one category of machine learning called *supervised learning*. As the name implies, the whole process of learning is designed like a teacher monitors the learning process. The learning process starts with an input dataset, which has a set of features or attributes along with the outputs mapped one to one. Here the input is called the *independent variables*, and the output is called the *dependent variables*. The values of output are called *labels*, which enable the training process to correlate with the input features easier. In short, the mapping function is generated from the dataset, which has known inputs and outputs along with the training process. The predicted output is generated through a mapping function that calculates a label for each set of inputs for which the outputs are unknown. This chapter describes the supervised learning algorithms in detail along with their simple implementations in Python as case study.

3.2 Regression

Regression is the data modeling that decides the best-fit line that covers all the data points plotted with independent variables against the dependent variables. The line that is best fit is determined by computing the distance between the possible set of lines and the data points. The line that has the smallest value of the sum of distances from all the data points is designated as the best-fit line. Various types of regression techniques exist and differ based on the form of regression lines and number of independent variables and type of dependent variable. Obviously, the nature of the data highly influences the type of regression to use.

3.2.1 Linear Regression

Linear regression is the basic type of regression that uncovers the linear relationship between the independent and dependent variable and is based on the following expression:

$$Y = \alpha_0 + \alpha_1 X_1 + \alpha_2 X_2 + \ldots\ldots + \alpha_n X_n + \epsilon \qquad \text{------ Eqn 1}$$

where Y depicts the predicted value of the dependent variable, α_0 is the intercept and value of Y when X is zero, α_i is the regression coefficient in deciding the Y value when X_i changes, and ϵ is the error in the regression.

$X_1, X_2, \ldots X_n$ are the independent variables, and if only one independent variable is used in deciding the values of dependent variable, then it is called *simple linear regression*. If more than one independent variable is used in deciding the value of the dependent variable, then it is called *multiple linear regression*.

3.2.2 Polynomial Regression

The regression technique uncovers the nonlinear relationship between the dependent and independent variables and is called *polynomial regression*. The linear model is used as the initial estimator and gradually fits on a curved or curvilinear line with nth degree polynomial equation. The polynomial equation is as follows:

$$Y = \alpha_0 + \alpha_1 X_1 + \alpha_2 X_1^2 + \ldots\ldots + \alpha_n X_1^n + \epsilon \qquad \text{------ Eqn 2}$$

where Y is the dependent or target variable, X_1 is the independent variable, $\alpha_0, \alpha_1, \alpha_2, ..., \alpha_n$ are the coefficients for 0, 1, 2, ..., n degree polynomial, and ϵ is error in the regression.

The regression analysis between the dependent and independent variables identifies the best-fit line. This polynomial regression is used to represent the nonlinear relationship between the dependent and independent variables more optimally than normal linear regression. Generally, Curved line is the best fit line for non-linear based relations in the datasets.

3.2.3 Bayesian Linear Regression

Linear regression in which the regression coefficients are determined using the Bayes theorem is referred to as Bayesian linear regression. The Bayes theorem states that

$$P(A/B) = \{P(B/A) . P(A)\} / P(B) \qquad \text{------ Eqn 3}$$

where P(A/B) is the probability of occurrence of event A when event B has occurred already, P(B/A) is the probability of occurrence of event B when event A has occurred already, and P(A) and P(B) are the probability of occurrence of events A and B.

In this regression, instead of finding the least squares, the posterior distribution of the features is calculated from the likelihood of the data, the prior probability of the features, and the normalization factor, based on the previous equation. Hence, the previous equation can be rewritten in regression as follows:

$$P (\beta|y, X) = \{P (y|\beta, X) * P(\beta|X)\} / P(y/X) \qquad \text{------ Eqn 4}$$

where $P(\beta|y,X)$ is the posterior distribution probability of the model parameters when the inputs and outputs are given, $P(y|\beta,X)$ denotes the likelihood of the data, $P(\beta|X)$ is the prior probability of the model parameters, and $P(y/X)$ is the normalization constant.

The probabilistic approach makes this regression type to generate accurate estimations for regression coefficients and hence is stronger than ordinary least square (OLS) linear regression.

3.2.4 Ridge Regression

Linear regression works well when the linear relationship between the input and target variables are stronger with appropriate regression coefficient values. But in real-time scenarios, the regression coefficients may have large or small values, which in turn make the model unstable and sensitive to input variable values. To recover the stability, regularization of regression coefficients is required.

Ridge regression is a type of regression that enables the models to use L2 regularization (sum of squares of coefficients multiplied with penalty term) for relatively larger or smaller regression coefficients values along with penalties added to the cost function. This shrinks the coefficients to reduce overfitting. Hence, it is called *penalized* or *regularized* linear regression.

From the linear regression section, the general linear equation for prediction is given as follows:

$$Y_{pred} = \alpha_0 + \alpha_1 X_1 + \alpha_2 X_2 + \ldots\ldots + \alpha_n X_n + \epsilon \qquad \text{------ Eqn 5}$$

The cost function for linear equation is as follows:

$$CF = \sum_{i=1}^{m}\left(Y_{actual} - Y_{pred}\right)^2 = \sum_{i=1}^{m}\left(Y_{actual} - \sum_{j=0}^{p}\alpha_j * x_{ij}\right)^2 \qquad \text{------ Eqn 6}$$

The cost function for ridge regression is as follows:

$$CF = \sum_{i=1}^{m}\left(Y_{actual} - Y_{pred}\right)^2 + penalty$$

$$= \sum_{i=1}^{m}\left(Y_{actual} - \sum_{j=0}^{p}\alpha_j * x_{ij}\right)^2 + \lambda\sum_{j=0}^{p}\alpha_j^2 \qquad \text{------ Eqn 7}$$

where λ is penalty term that is multiplied with the sum of squares of the coefficients and is known as L2 regularization.

3.2.5 Lasso Regression

Lasso regression is a type of regression like ridge regression to represent the nonlinear relationship of dependent and independent variable. Lasso regression is quite suitable for sparse (few nonzero values) models with data variables exhibiting a high level of multi-collinearity between them. The cost function of lasso regression is as follows:

$$CF = \sum_{i=1}^{m} \left(Y_{actual} - Y_{pred} \right)^2 + penalty$$

$$= \sum_{i=1}^{m} \left(Y_{actual} - \sum_{j=0}^{p} \alpha_j * x_{ij} \right)^2 + \lambda \sum_{j=0}^{p} \alpha_j \qquad \text{------ Eqn 8}$$

This is very similar to ridge regression except the penalty term (λ) is multiplied with the sum of coefficients simply instead of the squares of them.

LASSO stands for "Least Absolute Shrinkage and Selection Operator." Lasso employs shrinkage in which data converges toward the central point such as mean. Shrinkage is used to reduce the bias-variance trade-off and in turn reduces overfitting.

3.2.6 Case Study with Medical Applications

The following code blocks demonstrate each regression technique discussed earlier with the Python implementation of the dataset heart.csv available in the Kaggle dataset.

```
from google.colab import drive
drive.mount('/content/drive')

import os
os.chdir("/content/drive/MyDrive/Colab Notebooks/LAIoT")
```

############# **Linear Regression** #############

```python
import numpy as np
import pandas as pd
import matplotlib.pyplot as plt
from sklearn.model_selection import train_test_split
from sklearn.preprocessing import PolynomialFeatures
from sklearn.linear_model import LinearRegression
from sklearn.metrics import mean_squared_error, r2_score

# Load the dataset
df = pd.read_csv('/content/drive/MyDrive/Colab Notebooks/LAIoT/
heart.csv')

# Selecting features and target
y = df['thalach']
X = df[['age']]

# Train-test split
X_train, X_test, y_train, y_test = train_test_split(X, y, test_
size=0.3, random_state=0)

# Fit the Polynomial Regression model
lin_reg = LinearRegression()
lin_reg.fit(X_train, y_train)

# Model evaluation
y_pred_train = lin_reg.predict(X_train)
y_pred_test = lin_reg.predict(X_test)

train_rmse = np.sqrt(mean_squared_error(y_train, y_pred_train))
test_rmse = np.sqrt(mean_squared_error(y_test, y_pred_test))
train_r2 = r2_score(y_train, y_pred_train)
test_r2 = r2_score(y_test, y_pred_test)

print(f"Train RMSE: {train_rmse}")
```

```
print(f"Test RMSE: {test_rmse}")
print(f"Train R^2 Score: {train_r2}")
print(f"Test R^2 Score: {test_r2}")

# Visualize
plt.scatter(X, y, color='blue')
plt.scatter(X_train, y_train, color='red')
plt.plot(X_train, lin_reg.predict(X_train), color='green')
plt.title('Linear Regression')
plt.xlabel('Age')
plt.ylabel('Maximum Heart Rate')
plt.show()
```

Here's the output:

```
Train RMSE: 21.291457398658565
Test RMSE: 20.89197531256775
Train R^2 Score: 0.14825325239079368
Test R^2 Score: 0.16150667308285893
```

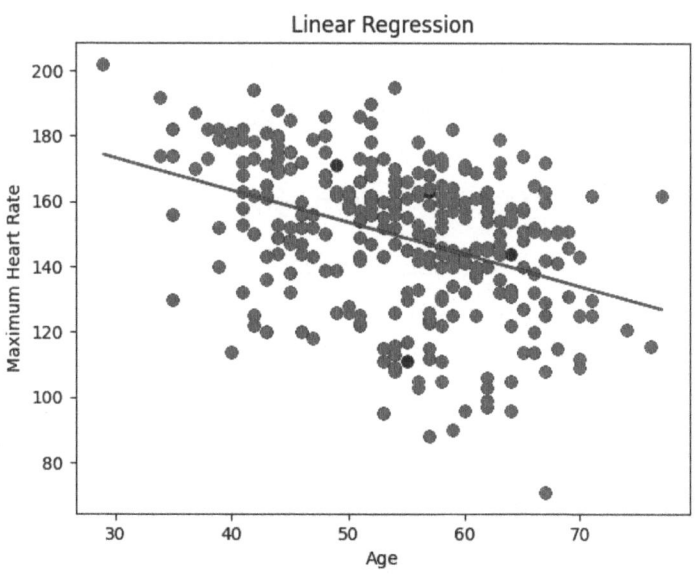

############# **Polynomial Regression** #############

```python
import numpy as np
import pandas as pd
import matplotlib.pyplot as plt
from sklearn.model_selection import train_test_split
from sklearn.preprocessing import PolynomialFeatures
from sklearn.linear_model import LinearRegression
from sklearn.metrics import mean_squared_error, r2_score

# Load the dataset
df = pd.read_csv('/content/drive/MyDrive/Colab Notebooks/LAIoT/
heart.csv')

# Selecting features and target
y = df['thalach']
X = df[['age']]

# Train-test split
X_train, X_test, y_train, y_test = train_test_split(X, y, test_
size=0.3, random_state=0)

# Apply Polynomial Features
poly_features = PolynomialFeatures(degree=2)
X_poly_train = poly_features.fit_transform(X_train)
X_poly_test = poly_features.transform(X_test)

# Fit the Polynomial Regression model
poly_reg = LinearRegression()
poly_reg.fit(X_poly_train, y_train)

# Model evaluation
y_pred_train = poly_reg.predict(X_poly_train)
y_pred_test = poly_reg.predict(X_poly_test)
```

```
train_rmse = np.sqrt(mean_squared_error(y_train, y_pred_train))
test_rmse = np.sqrt(mean_squared_error(y_test, y_pred_test))
train_r2 = r2_score(y_train, y_pred_train)
test_r2 = r2_score(y_test, y_pred_test)

print(f"Train RMSE: {train_rmse}")
print(f"Test RMSE: {test_rmse}")
print(f"Train R^2 Score: {train_r2}")
print(f"Test R^2 Score: {test_r2}")

# Visualize
plt.scatter(X, y, color='blue')
plt.scatter(X_train, y_train, color='red')
plt.plot(X_train, poly_reg.predict(poly_features.fit_
transform(X_train)), color='green')
plt.title('Polynomial Regression')
plt.xlabel('Age')
plt.ylabel('Maximum Heart Rate')
plt.show()
```

Here's the output:

```
Train RMSE: 21.246248868365058
Test RMSE: 20.897371225131927
Train R^2 Score: 0.1518664703544036
Test R^2 Score: 0.16107349040017727
```

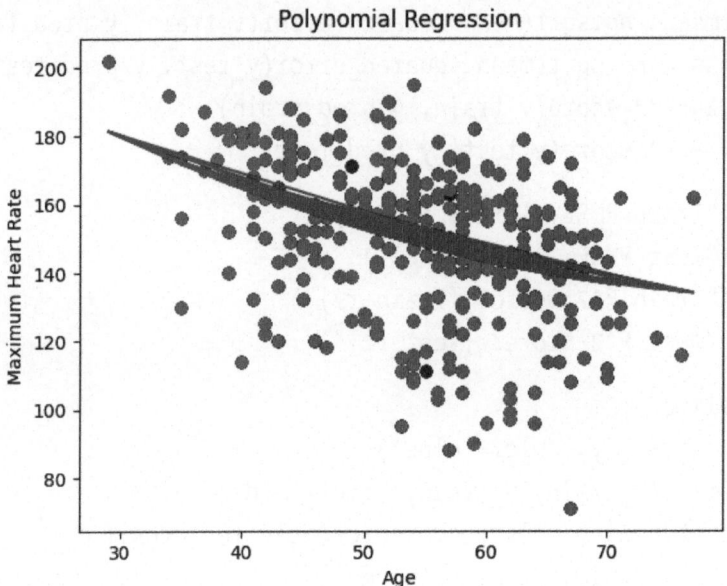

############# **Bayesian Regression** #############

```
import numpy as np
import pandas as pd
import matplotlib.pyplot as plt
from sklearn.model_selection import train_test_split
from sklearn.linear_model import BayesianRidge
from sklearn.metrics import mean_squared_error, r2_score

# Load the dataset
df = pd.read_csv('/content/drive/MyDrive/Colab Notebooks/LAIoT/
heart.csv')

# Selecting features and target
y = df['thalach']
X = df[['age']]
```

```python
# Train-test split
X_train, X_test, y_train, y_test = train_test_split(X, y, test_
size=0.3, random_state=0)

# Fit the Bayesian Regression model
bayesian_reg = BayesianRidge()
bayesian_reg.fit(X_train, y_train)

# Model evaluation
y_pred_train = bayesian_reg.predict(X_train)
y_pred_test = bayesian_reg.predict(X_test)

train_rmse = np.sqrt(mean_squared_error(y_train, y_pred_train))
test_rmse = np.sqrt(mean_squared_error(y_test, y_pred_test))
train_r2 = r2_score(y_train, y_pred_train)
test_r2 = r2_score(y_test, y_pred_test)

print(f"Train RMSE: {train_rmse}")
print(f"Test RMSE: {test_rmse}")
print(f"Train R^2 Score: {train_r2}")
print(f"Test R^2 Score: {test_r2}")

# Visualize
plt.scatter(X, y, color='blue')
plt.scatter(X_train, y_train, color='red')
plt.plot(X_train, bayesian_reg.predict(X_train), color='green')
plt.title('Bayesian Regression')
plt.xlabel('Age')
plt.ylabel('Maximum Heart Rate')
plt.show()
```

Here's the output:

```
Train RMSE: 21.291576704435904
Test RMSE: 20.892545623905956
Train R^2 Score: 0.14824370691059263
Test R^2 Score: 0.16146089390006724
```

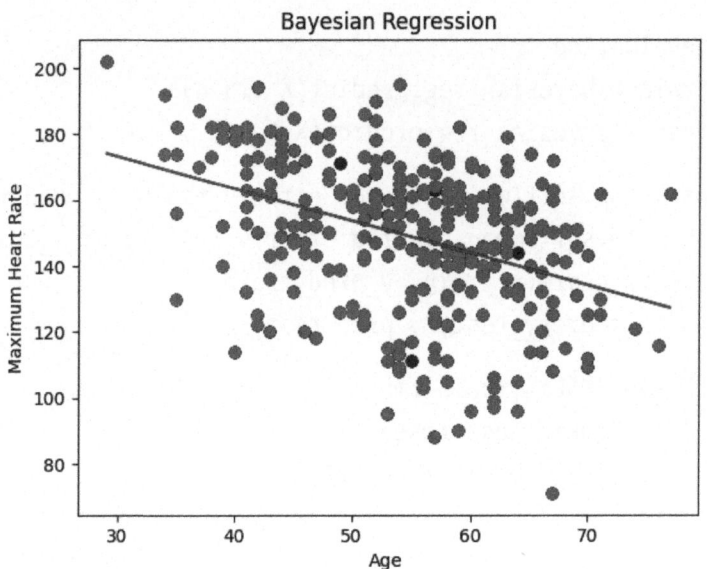

############# **Lasso Regression** #############

```
import numpy as np
import pandas as pd
import matplotlib.pyplot as plt
from sklearn.model_selection import train_test_split
from sklearn.linear_model import Lasso
from sklearn.metrics import mean_squared_error, r2_score

# Load the dataset
df = pd.read_csv('/content/drive/MyDrive/Colab Notebooks/LAIoT/
heart.csv')
```

```python
# Selecting features and target
y = df['thalach']
X = df[['age']]

# Train-test split
X_train, X_test, y_train, y_test = train_test_split(X, y,
test_size=0.3, random_state=0)

# Fit the Lasso Regression model
lasso_reg = Lasso(alpha=0.1)  # You can adjust the alpha
(regularization strength)
lasso_reg.fit(X_train, y_train)

# Model evaluation
y_pred_train = lasso_reg.predict(X_train)
y_pred_test = lasso_reg.predict(X_test)

train_rmse = np.sqrt(mean_squared_error(y_train, y_pred_train))
test_rmse = np.sqrt(mean_squared_error(y_test, y_pred_test))
train_r2 = r2_score(y_train, y_pred_train)
test_r2 = r2_score(y_test, y_pred_test)

print(f"Train RMSE: {train_rmse}")
print(f"Test RMSE: {test_rmse}")
print(f"Train R^2 Score: {train_r2}")
print(f"Test R^2 Score: {test_r2}")

# Visualize
plt.scatter(X, y, color='blue')
plt.scatter(X_train, y_train, color='red')
plt.plot(X_train, lasso_reg.predict(X_train), color='green')
plt.title('Lasso Regression')
plt.xlabel('Age')
plt.ylabel('Maximum Heart Rate')
plt.show()
```

Here's the output:

```
Train RMSE: 21.291460286914507
Test RMSE: 20.8920474711158
Train R^2 Score: 0.14825302130630535
Test R^2 Score: 0.16150088094835757
```

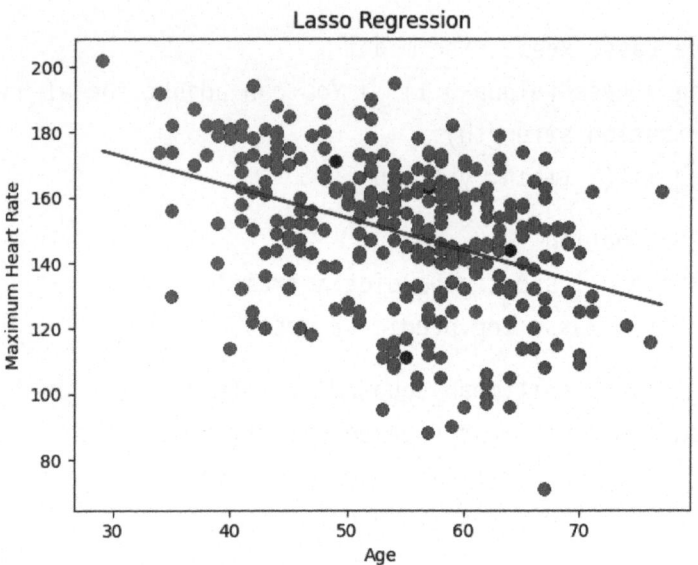

############# **Ridge Regression** #############

```
import numpy as np
import pandas as pd
import matplotlib.pyplot as plt
from sklearn.model_selection import train_test_split
from sklearn.linear_model import Ridge
from sklearn.metrics import mean_squared_error, r2_score

# Load the dataset
df = pd.read_csv('/content/drive/MyDrive/Colab Notebooks/LAIoT/
heart.csv')
```

```python
# Selecting features and target
y = df['thalach']
X = df[['age']]

# Train-test split
X_train, X_test, y_train, y_test = train_test_split(X, y,
test_size=0.3, random_state=0)

# Fit the Ridge Regression model
ridge_reg = Ridge(alpha=1.0)  # You can adjust the alpha
(regularization strength)
ridge_reg.fit(X_train, y_train)

# Model evaluation
y_pred_train = ridge_reg.predict(X_train)
y_pred_test = ridge_reg.predict(X_test)

train_rmse = np.sqrt(mean_squared_error(y_train, y_pred_train))
test_rmse = np.sqrt(mean_squared_error(y_test, y_pred_test))
train_r2 = r2_score(y_train, y_pred_train)
test_r2 = r2_score(y_test, y_pred_test)

print(f"Train RMSE: {train_rmse}")
print(f"Test RMSE: {test_rmse}")
print(f"Train R^2 Score: {train_r2}")
print(f"Test R^2 Score: {test_r2}")

# Visualize
plt.scatter(X, y, color='blue')
plt.scatter(X_train, y_train, color='red')
plt.plot(X_train, ridge_reg.predict(X_train), color='green')
plt.title('Ridge Regression')
plt.xlabel('Age')
plt.ylabel('Maximum Heart Rate')
plt.show()
```

Here's the output:

```
Train RMSE: 21.291457399203768
Test RMSE: 20.891976262576687
Train R^2 Score: 0.1482532523471729
Test R^2 Score: 0.16150659682619484
```

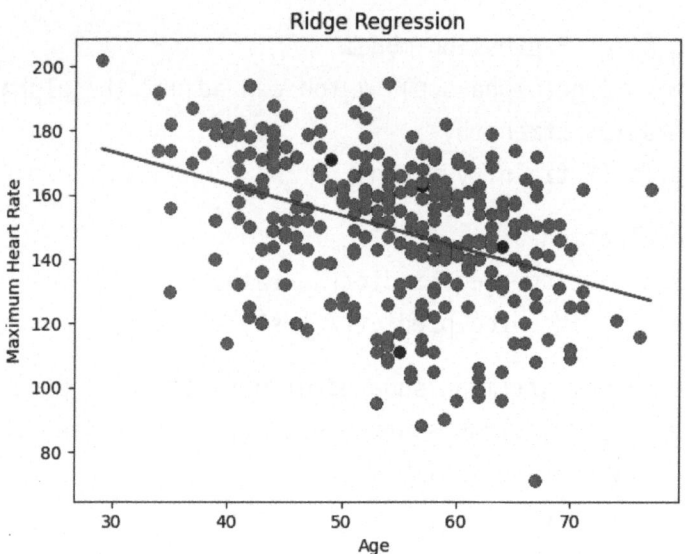

3.3 Classification

The prediction of labels is called *classification* when the target variable is
not continuous in nature. Classification is the process of understanding,
recognizing, and categorizing the data into an existing category. It is a
supervised machine learning technique where the models are trained to
predict the label of the given data. It works based on the probability or
threshold score to determine the category of the data. It trains the model
with the training set of data and evaluates with the test set of data. Then
the model is subjected to the prediction of new data. Classification works
well on both structured and unstructured data.

3.3.1 Logistic Regression

Linear regression works on data that is continuous in nature. Instead, logistic regression is based on the probability of events. Hence, the equation of linear type needs to be changed to accommodate predicting the probability with arbitrary distributions, not only normal distribution. The predictions in classification are not normally distributed, and it can be larger than 1 and lesser than 0.

The logistic regression equation is derived as follows:

The odds of an event when the probability of success is P is given by this:

$$\text{Odds} = P/1\text{-}P \qquad \text{------ Eqn 9}$$

The linear function is given by $Y = \alpha_0 + \alpha_1 X_1$.

This can be transformed into logistic function as follows:

$$\ln(P/1\text{-}P) = \alpha_0 + \alpha_1 X_1 \qquad \text{------ Eqn 10}$$

Exponentiating both sides results in the following:

$$\frac{P}{1-P} = e^{\alpha_0 + \alpha_1 X_1} \qquad \text{------ Eqn 11}$$

Then to find P, use this:

$$P = \frac{e^{\alpha_0 + \alpha_1 X_1}}{1 + e^{\alpha_0 + \alpha_1 X_1}} \qquad \text{------ Eqn 12}$$

The equation of sigmoid function is as follows:

$$P(x) = \frac{1}{1 + e^{-(\alpha_0 + \alpha_1 X_1)}} \qquad \text{------ Eqn 13}$$

Hence, the logistic regression works based on the previous sigmoid equation.

3.3.2 Decision Trees

Decision trees is a classification method based on the hierarchical tree structure. It uses the conditional control constructs, and it is a nonparametric approach. The decision tree is constructed based on the decisions along with chances of events, expense of resources, and utilization to generate their potential outcomes. It is accomplished with the various components such as root node, branches, leaf nodes, subtree, pruning, and decision nodes.

Root node depicts the entire population at which the dataset starts segregating based on its features and values. Decision nodes or intermediate nodes in the decision tree represent intermediate conditions and separation of branches. A subtree is a subsection of a decision tree, and it represents part of the tree based on a specific condition. Branches are a new division, which leads to subtrees. Pruning is the process of eliminating the nodes to avoid overfitting and simplifying the generation of model. Finally, leaf nodes represent the classification outcome and indicate the category of the data; they are also called *terminal nodes*.

Decision trees work based on entropy, which indicates the uncertainty in the dataset. The formula for calculating entropy is given as follows:

$$E(S) = - p_+ \log p_+ - p_- \log p_- \qquad \text{------ Eqn 14}$$

where p_+ is the probability of + (positive) class, p_- is the probability of – (negative) class, and S is a subset in training example.

3.3.3 Naïve Bayes

The naïve Bayes classifier works based on the assumption that the presence of a particular feature in a class is independent and unrelated to the existence of any other feature. It is based on probability and applies the Bayes theorem.

The Bayes theorem states that

$$P(A/B) = \{P(B/A) . P(A)\} / P(B)$$

where $P(A/B)$ is the posterior probability of class A (target) when given predictor B attributes; $P(B/A)$ is the likelihood, which is the probability of the predictor given class A; $P(A)$ is the prior probability of class A; and $P(B)$ is the prior probability of class B.

The naïve Bayes algorithm works with the following steps:

1. Construct the frequency table from the dataset. The frequency table summarizes the happening and not happening of the given class (A as in equation) under various conditions (attributes).

2. Build the likelihood table with the probabilities determined for each attribute for the happening or occurring of the class (A as target).

3. The posterior probability is determined with the naïve Bayes equation mentioned earlier for each class. The class that has the highest posterior probability is designated as the predicted class.

3.3.4 Random Forest

Random forest is an ensemble technique for classification, which works based on generating multiple decision trees during the training phase. The output is the maximum voted class or category to determine the final value for classification. It also works for regression in which the average is computed as the final output.

The combination of more than one model is called the *ensemble*, and it is achieved with two different methods called *bagging* and *boosting*. Bagging is selecting different subsets from the dataset with replacement for training, and majority votes is the final output. Boosting is building a sequential model, which combines the weak learning models with strong ones in order to get high accuracy in the final model.

The best part of this method is the generation of the number of trees (not only the decision tree, but also supports other methods) with different subset of features from the dataset. Features are selected randomly and best among them is searched. From this process, the set of trees generated denotes the wide diversity of feature selection to yield better overall results.

3.3.5 Support Vector Machines

Support vector machine (SVM) is a supervised algorithm that is more suitable for classification than regression though it works for both. It works well for complex and smaller datasets. SVM must identify the decision boundaries or hyperplane for separating the different categories of data. There exists more than one hyperplane to separate the categories of data, as shown in Figure 3-1. But only one will be the largest and has the maximum margin compared to the other hyperplanes.

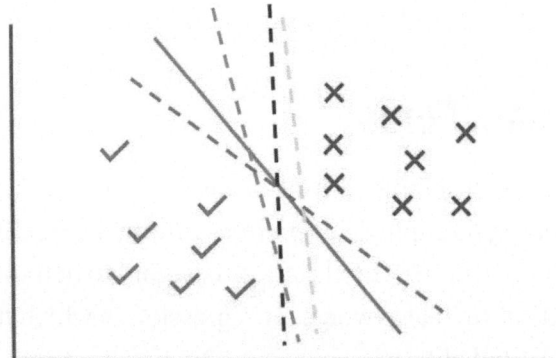

Figure 3-1. *SVM, sample set of possible hyperplanes*

The margin is the distance between the support vectors from each class of data, as shown in Figure 3-2. The maximization of margin distance is recommended to classify the future data points with more confidence.

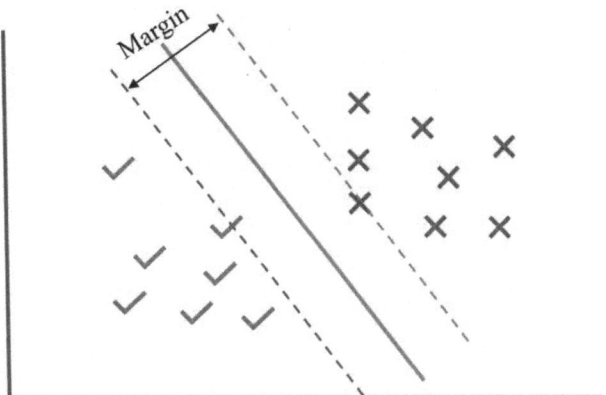

Figure 3-2. *SVM, hyperplane and its support vector with maximum margin*

3.3.6 Case Study with Agriculture

The following code blocks demonstrates each classification technique discussed earlier with the Python implementation of the dataset agriculture.csv available in the Kaggle dataset.

```
from google.colab import drive
drive.mount('/content/drive')
import os
os.chdir("/content/drive/MyDrive/Colab Notebooks/LAIoT")
```

#############**Logistic Regression** #################

```
import pandas as pd
from sklearn.cluster import KMeans
import matplotlib.pyplot as plt
from sklearn import metrics
from sklearn.model_selection import train_test_split
from sklearn.linear_model import LogisticRegression
from sklearn.metrics import confusion_matrix, accuracy_score
import numpy as np
```

```
# Load the dataset
df = pd.read_csv('agriculture.csv')

# Selecting features and target
df['Number_Weeks_Used'] = df['Number_Weeks_Used'].
fillna(df['Number_Weeks_Used'].mean())
X = df[['Estimated_Insects_Count','Crop_Type','Soil_
Type','Pesticide_Use_Category', 'Number_Doses_Week',
'Number_Weeks_Used','Number_Weeks_Quit','Season']].values
Y = df[['Crop_Damage']].values

# Train-test split
X_train, X_test,y_train,y_test = train_test_split(X,Y,
test_size=0.3)

#Fit the Logistic Regression model
lr = LogisticRegression(solver='lbfgs', max_iter=5000)
lr.fit(X_train,y_train)

#Model Evaluation
pred = lr.predict(X_test)
ac = accuracy_score(y_test,pred)
cm = confusion_matrix(y_test, pred)
print(ac)
print(cm)

#Visualization of metric
cm_display = metrics.ConfusionMatrixDisplay(confusion_matrix =
cm, display_labels = ["alive", "othercause","pesticides"])
cm_display.plot()
plt.show()
```

Here's the output:

```
Accuracy = 0.8303323580163553
Confusion Matrix =
 [[21948    231      0]
 [ 3600    187      0]
 [  659     33      0]]
```

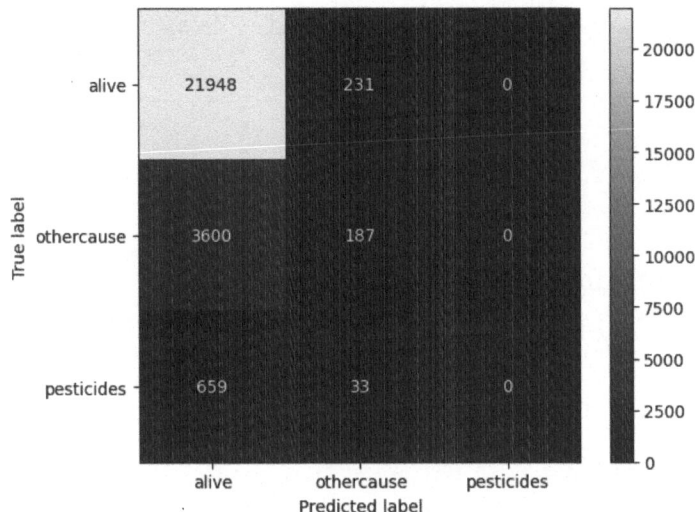

############Decision Trees####################

```
import pandas as pd
from sklearn.cluster import KMeans
import matplotlib.pyplot as plt
from sklearn import metrics
from sklearn.model_selection import train_test_split
from sklearn.tree import DecisionTreeClassifier
from sklearn.metrics import confusion_matrix, accuracy_score
import numpy as np
```

```
# Load the dataset
df = pd.read_csv('agriculture.csv')

# Selecting features and target
df['Number_Weeks_Used'] = df['Number_Weeks_Used'].
fillna(df['Number_Weeks_Used'].mean())
X = df[['Estimated_Insects_Count','Crop_Type','Soil_
Type','Pesticide_Use_Category', 'Number_Doses_Week',
'Number_Weeks_Used','Number_Weeks_Quit','Season']].values
Y = df[['Crop_Damage']].values

# Train-test split
X_train, X_test,y_train,y_test = train_test_split(X,Y,test_
size=0.3)

#Fit the Decision Tree Regression model
agriTree = DecisionTreeClassifier(criterion="entropy",
max_depth = 4)
agriTree.fit(X_train,y_train)

#Model Evaluation
y_pred = agriTree.predict(X_test)
ac = accuracy_score(y_test,y_pred)
cm = confusion_matrix(y_test, y_pred)
print(ac)
print(cm)

#Visualization of metric
cm_display = metrics.ConfusionMatrixDisplay(confusion_matrix =
cm, display_labels = ["alive", "othercause","pesticides"])
cm_display.plot()
plt.show()
```

Here's the output:

```
Accuracy = 0.8383974791807337
Confusion Matrix =
 [[22178    39     0]
 [ 3596   172     0]
 [  628    45     0]]
```

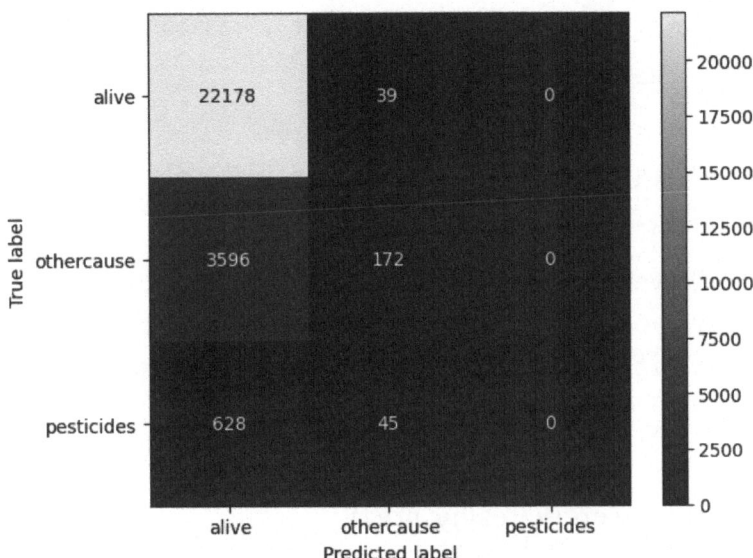

###############Naive Bayes#####################

```
import pandas as pd
from sklearn.cluster import KMeans
import matplotlib.pyplot as plt
from sklearn import metrics
from sklearn.model_selection import train_test_split
from sklearn.naive_bayes import GaussianNB
from sklearn.metrics import confusion_matrix, accuracy_score
import numpy as np
```

```
# Load the dataset
df = pd.read_csv('agriculture.csv')

# Selecting features and target
df['Number_Weeks_Used'] = df['Number_Weeks_Used'].
fillna(df['Number_Weeks_Used'].mean())
X = df[['Estimated_Insects_Count','Crop_Type','Soil_
Type','Pesticide_Use_Category', 'Number_Doses_Week',
'Number_Weeks_Used','Number_Weeks_Quit','Season']].values
Y = df[['Crop_Damage']].values

# Train-test split
X_train, X_test,y_train,y_test = train_test_split(X,Y,test_
size=0.3)

#Fit the NaiveBayes Regression model
classifier = GaussianNB()
classifier.fit(X_train, y_train)

#Model Evaluation
y_pred = classifier.predict(X_test)
ac = accuracy_score(y_test,y_pred)
cm = confusion_matrix(y_test, y_pred)
print(ac)
print(cm)

#Visualization of metric
cm_display = metrics.ConfusionMatrixDisplay(confusion_matrix =
cm, display_labels = ["alive", "othercause","pesticides"])
cm_display.plot()
plt.show()
```

Here's the output:

```
Accuracy = 0.817278115387501
Confusion Matrix =
 [[20972    1175       1]
  [ 3006     815       0]
  [  504     185       0]]
```

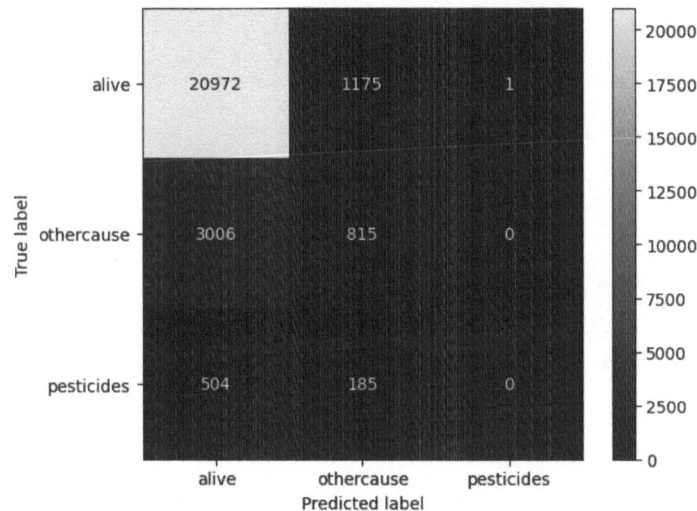

################Random Forest##############

```
import pandas as pd
from sklearn.cluster import KMeans
import matplotlib.pyplot as plt
from sklearn import metrics
from sklearn.model_selection import train_test_split
from sklearn.ensemble import RandomForestClassifier
from sklearn.metrics import confusion_matrix, accuracy_score
import numpy as np
```

```python
# Load the dataset
df = pd.read_csv('agriculture.csv')

# Selecting features and target
df['Number_Weeks_Used'] = df['Number_Weeks_Used'].
fillna(df['Number_Weeks_Used'].mean())
X = df[['Estimated_Insects_Count','Crop_Type','Soil_
Type','Pesticide_Use_Category', 'Number_Doses_Week',
'Number_Weeks_Used','Number_Weeks_Quit','Season']].values
Y = df[['Crop_Damage']].values

# Train-test split
X_train, X_test,y_train,y_test = train_test_split(X,Y,
test_size=0.3)

#Fit the Random Forest Regression model
rf_model = RandomForestClassifier(n_estimators=50,
max_features="sqrt",random_state=44)
rf_model.fit(X_train,y_train)

#Model Evaluation
y_pred = rf_model.predict(X_test)
ac = accuracy_score(y_test,y_pred)
cm = confusion_matrix(y_test, y_pred)
print(ac)
print(cm)

#Visualization of metric
cm_display = metrics.ConfusionMatrixDisplay(confusion_matrix =
cm, display_labels = ["alive", "othercause","pesticides"])
cm_display.plot()
plt.show()
```

Here's the output:

```
Accuracy = 0.8209918223422612
Confusion Matrix =
 [[21088   1011    114]
 [ 2835    764    139]
 [  469    204     34]]
```

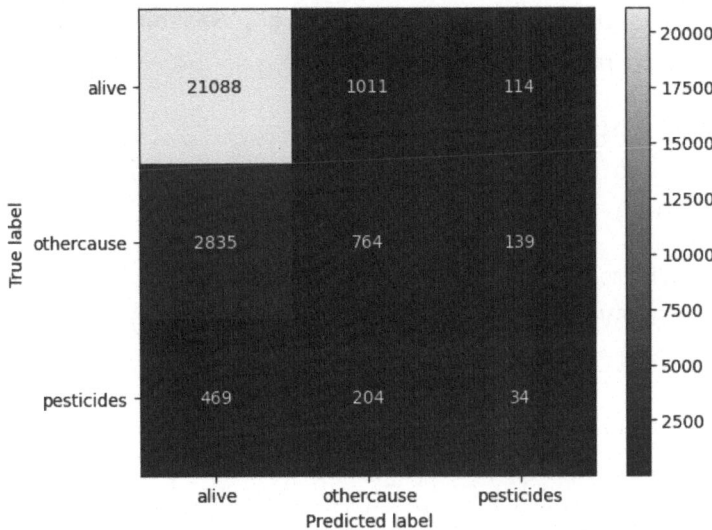

############Support Vector Machine #################

```
import pandas as pd
from sklearn.cluster import KMeans
import matplotlib.pyplot as plt
from sklearn import metrics
from sklearn.model_selection import train_test_split
from sklearn.svm import SVC
from sklearn.metrics import confusion_matrix, accuracy_score
import numpy as np
```

105

```python
# Load the dataset
df = pd.read_csv('agriculture.csv')

# Selecting features and target
df['Number_Weeks_Used'] = df['Number_Weeks_Used'].
fillna(df['Number_Weeks_Used'].mean())
X = df[['Estimated_Insects_Count','Crop_Type','Soil_
Type','Pesticide_Use_Category', 'Number_Doses_Week',
'Number_Weeks_Used','Number_Weeks_Quit','Season']].values
Y = df[['Crop_Damage']].values

#Train-test Split
X_train, X_test,y_train,y_test = train_test_split(X,Y,
test_size=0.3)

#Fit the SVM model
classifier = SVC(kernel='rbf', random_state = 1)
classifier.fit(X_train, y_train)

#Model Evaluation
y_pred = classifier.predict(X_test)
ac = accuracy_score(y_test,y_pred)
cm = confusion_matrix(y_test, y_pred)
print(ac)
print(cm)

#Visualization of metric
cm_display = metrics.ConfusionMatrixDisplay(confusion_matrix =
cm, display_labels = ["alive", "othercause","pesticides"])
cm_display.plot()
plt.show()
```

Here's the output:

```
Accuracy = 0.8382474304148848
Confusion Matrix =
 [[22346     0      0]
  [ 3606     0      0]
  [  706     0      0]]
```

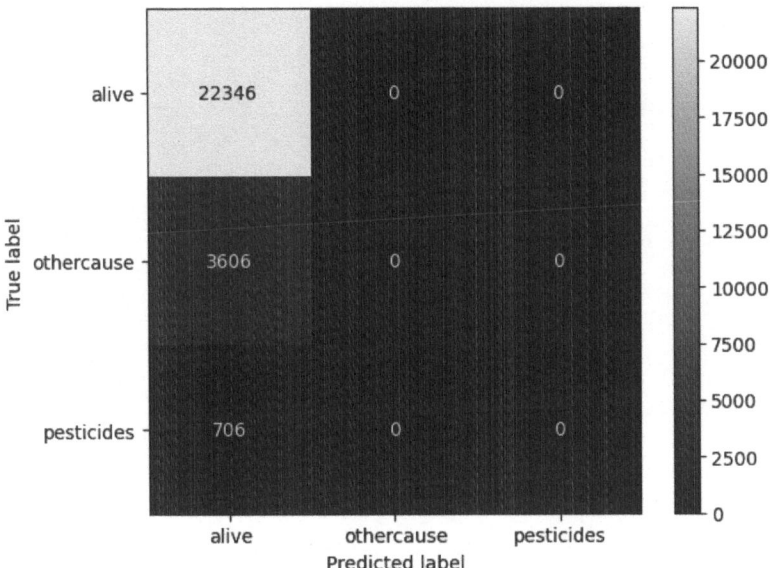

CHAPTER 4

Unsupervised Algorithms

4.1 Introduction

In real time, it is hard to generate data in an unstructured format and deduce insights from it. Unsupervised learning determines the relationship between the data or group of data points. It uses the features of data and not the label of data to study patterns, and hence experts compare it with human intellect. The unstructured data is converted to structured data by identifying similarities between the individual data records in the dataset using unsupervised learning.

Unsupervised learning is more advantageous in handling complex data, deriving insights from raw data without the requirement of labels, and recognizing the patterns involved between unstructured data in real time; it is also cheaper than supervised learning. Though it has many benefits, there are challenges involved. It consumes more time for training, it is difficult to identify the hidden patterns, and the results are unpredictable as there are no labels in the dataset to verify.

This chapter explores the various unsupervised learning techniques such as K-means clustering, hierarchical clustering, principal component analysis, independent component analysis, anomaly detection, neural networks, the apriori algorithm, and singular value decomposition, along with their Python implementations.

4.2 K-Means Clustering

K-means clustering is a learning algorithm used to solve the problem of grouping data in the dataset into different clusters. It is generally used in machine learning and data science applications. It is similar in concept to the K-NN (nearest neighbor) technique in supervised learning used for regression and classification. K-means clustering separates the entire dataset into nonoverlapping clusters based on the centroids, as shown in Figure 4-1. The sum of the distances between each data point and its cluster centroid needs to be minimized to add that data point in the corresponding cluster. Data with similar features belongs to the same cluster. Either the cluster size K or number of clusters is predefined based on the requirement of application, or an optimized K value will be decided by iterating the algorithm with a different K value. The formula for finding the objective function is as follows:

$$\sum_{i=1}^{m}\left(\sum_{k=1}^{K}\left(W_{ik}*\|x_i - \mu_k\|^2\right)\right) \qquad ----- \text{Eqn 1}$$

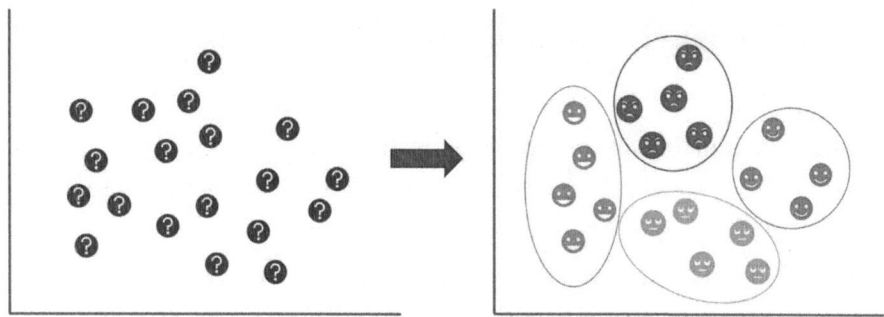

Figure 4-1. *Initial and final iteration of K-means clustering visualization*

The algorithm steps are as follows:

1. Identify the optimized K value to determine the number of clusters.

2. Initially choose random K points as centroids. (It is not necessarily in the input dataset.)

3. Designate the cluster to each data point by calculating the distance between the data point and the K centroids and finding the closest cluster with minimum distance.

4. Compute the variation in the centroids chosen earlier and identify a new centroid for each cluster.

5. Repeat step 3 to assign a new cluster if required based on new centroids computed in step 4.

6. If any reassignment of a cluster happens, then go to step 4. Otherwise, the model is complete and ready.

4.3 Hierarchical Clustering

In general, individual records in the dataset are examined and grouped into different clusters. In hierarchical clustering, the grouping of clusters is performed at the next higher level, and the grouping continues in a bottom-up approach until it reaches a stage in which only one cluster depicts all the records in the dataset. The grouping in various levels is represented as a tree in a top-down approach with each branch representing a more specific subcluster based on conditions on the value range of the attributes in the dataset.

The representation of clusters as tree shaped in hierarchical clustering is called a *dendrogram*. The clusters resulting from using K-means and hierarchical clustering might be the same for a few datasets though they are different conceptually in generating clusters. The role of hierarchical clustering is needed in applications that do not dictate a predefined number of clusters with the same size.

Hierarchical clustering is classified into two different approaches, agglomerative and divisive.

Agglomerative clustering is the method of generating clusters in a bottom-up approach by starting from an individual dataset as single clusters. Next, it combines the closely related cluster datasets together to form next-level clusters. Combining a cluster at each level from the bottom is also called *additive hierarchical* clustering.

Divisive hierarchical clustering works in complement to agglomerative. It works in a top-down approach starting from a single cluster. Later at each level, the cluster will divide further based on the threshold or range of value of the data points. This will be repeated until each cluster contains only a single data point. It is also called *divisive hierarchical clustering*.

The data points in the dataset are grouped into clusters based on the distance measure. The distance measure can be any one of the following between the data points A and B:

- **Euclidean distance measure:** Euclidean distance represents the length of the straight line connecting data points A and B. It is computed as the sum of squares of each dimension distance and then the square root of it.

$$D_E = \sqrt{\sum_{i=1}^{n}\left(A_i - B_i\right)^2} \qquad \text{------- Eqn 2}$$

- **Squared Euclidean distance measure:** This distance measure finds the sum of squares of each dimension distance as the measurement between the two points.

$$D_{SE} = \sum_{i=1}^{n}\left(A_i - B_i\right)^2 \qquad \text{------- Eqn 3}$$

- **Manhattan distance measure:** Manhattan distance computes the distance as the sum of horizontal and vertical directions or distances in the direction of axes at right angles.

$$D_M = \sum_{i=1}^{n}\left|A_{ix} - B_{ix}\right| + \left|A_{iy} - B_{iy}\right| \qquad \text{------- Eqn 4}$$

- **Cosine distance measure:** The cosine distance measure finds the angle between the vector data points.

$$D_C = \frac{\sum_{i=0}^{n-1}\left(A_i - B_i\right)}{\sum_{i=0}^{n-1} A_i^2 \ X \ \sum_{i=0}^{n-1} B_i^2} \qquad \text{------ Eqn 5}$$

The greater the angle between the data points, the farther the data points are. This measure is similar to Euclidean distance and results in similar outcomes, whereas Manhattan method produces different outcomes.

4.4 Principal Component Analysis

Generally, a dataset provided for learning is large enough to train the model in an appropriate manner to get good accuracy. But not all variables contribute to better analysis, which can be a waste of computation time and resources anyway. Principal component analysis (PCA) is an unsupervised learning technique used to eliminate the less critical variables without losing information for analysis. It is also called the *dimensionality reduction method* in machine learning.

The statistical process of orthogonal transformation is used to convert the correlated features to linearly uncorrelated features. These newly converted features are known as *principal components*, which enable the exploratory analysis of data in predictive algorithms. PCA focuses the variance in each attribute to infer the segregation of classes and in turn reduce the dimensionality. It is a feature extraction technique that precisely chooses key variables and excludes the less important variables.

The PCA algorithm is as follows:

1. **Organizing the dataset:** The dataset is represented as 2D structure with rows denoting the data items, columns denoting the features, and the number of columns denoting the dimension of dataset. The dataset is separated into two parts: the training set and the validation set.

2. **Examining and standardizing the data:** The data needs to be normalized by manipulating the mean, variance, or standard deviation, whichever is suitable based on the nature of the corresponding feature, and standardizing it.

3. **Computing the covariance matrix:** The covariance of each feature with itself is its variance, and it occupies the diagonal of the matrix. The covariance of two features is commutative, and hence the matrix is symmetric in nature with respect to its diagonal.

4. **Calculating eigen values, eigen vectors, and its normalization:** The eigen values of n × n matrix M is denoted by $\lambda_1, \lambda_2, \lambda_3, ..., \lambda_n$ and computed as |M- λI| = 0 where I is the identity matrix. The eigen vectors $e_1, e_2, e_3, ...e_n$ are computed from (M- λ_iI)e_i = 0. The eigen vector is normalized by dividing each element of eigen vector by the square root of sum of squares of each element in the eigen vector.

5. **Determining the principal components (PCs):** The sorted normalized vector in decreasing order is multiplied with a covariance matrix to get the principal components.

6. **Removing less important features:** It is inferred from the resulting PCs that less important variables whose contribution for prediction is negligible can be removed to form a new dataset.

4.5 Independent Component Analysis

Independent component analysis (ICA) is an unsupervised learning technique to isolate the sources of mixed signals. The independent components of the mixed signals are focused and separated in ICA, unlike PCA where the variance between data points is the focus. For example, more than one sound source, as shown in Figure 4-2, is recorded through a microphone, and the resultant sound wave is the mixed signal. The independent components are the sound generated from the radio, cow, bird, man, and kid. The sound needs to be separated from the mixed signal.

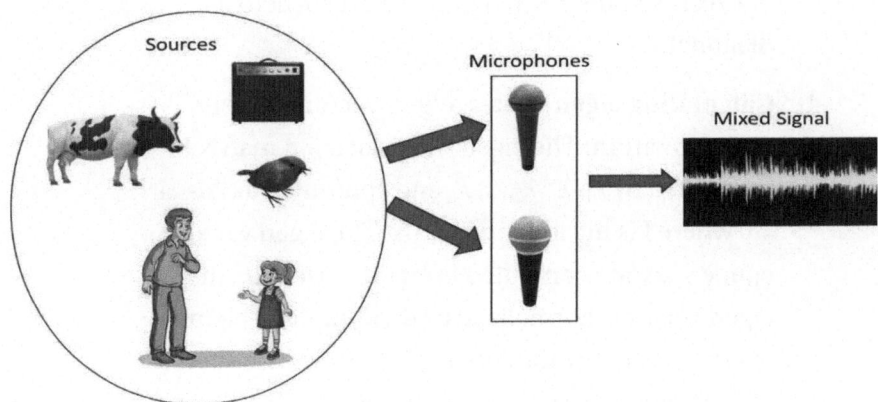

Figure 4-2. *Multiple sources of sound recorded in two microphones as a mixed signal*

The generation of mixing signal is defined mathematically.

Let independent source signals be $S = (s_1 \ s_2 \ s_3 \ s_4 \ s_5)$. The mixed signals are $x_1 = a_{11} \ s_1 + a_{12} \ s_2 + a_{13} \ s_3 + a_{14} \ s_4 + a_{15} \ s_5$ and $x_2 = a_{21} \ s_1 + a_{22} \ s_2 + a_{23} \ s_3 + a_{24} \ s_4 + a_{25} \ s_5$, where $A = (a_{11} \ a_{12} \ a_{13} \ a_{21} \ a_{22} \ a_{23} \ \ a_{14} \ a_{15} \ a_{24} \ a_{25})$ is the mixing coefficient.

The basic assumptions in ICA are that the measured signal is the linear combination of sources, the source signals are independent statistically, and each source signal value follows a non-Gaussian distribution. But at the same time, mixed signals are Gaussian, nonindependent, and complex.

The procedure of ICA is as follows.

The mixed signal is transformed into the separated source signals y1, y2, y3, y4, and y5 with the use of unmixing signals x1 and x2. Let $y_1 = ax_1 + bx_2$, $y_2 = cx_1 + dx_2$, ... until $y_5 = ix_1 + jx_2$, where $U = (a \ b \ c \ d \ e \ f \ g \ h \ i \ j)$

is the unmixing coefficients. The source signal y_1 is extracted from the mixed signal using unmixing coefficients (a, b). The unmixing coefficients denotes a point with the weight vector as $u1 = (a,b)^T$ to extract y_1. The unmixing coefficients of all the source signals can be written as the weight matrix $U = (u_1\ u_2\ u_3\ u_4\ u_5)^T$. The unmixing process of y_1 is performed by rotating the rows in U until the determination of the orientation that is orthogonal to all other transformed signals except y1 to extract it. Similarly, other signals are extracted with corresponding coefficients and determining the orientation suitable to extract each source signal. During this unmixing process, the length and orientation have an impact on the extracted signal, and hence extracted signals may not be the same but similar to the original source signals.

The preprocessing steps involved in ICA are centering and whitening. The centering process is to subtract the mean value from the mixed signals. It is denoted as $C = X-\mu$ where C is the mixed signal after the centering process, X is the original mixed signal, and μ is the mean value. The whitening process is further divided into two steps: decorrelation and scaling. The decorrelation makes all the signals uncorrelated to each other with their covariance as zero. Each signal after being uncorrelated is scaled to be with unit variance. After this step, the data becomes a sphere denoting it to be rotationally symmetric. Hence, this process is also called *sphering*. The various types of ICA algorithm are FastICA, which extracts signals by maximizing non-Gaussianity with negentropy; Infomax, which extracts signals by maximizing the mutual information between mixed signals and independent components; and Joint Approximated Diagonalization of Eigenmatrices (JADE), which extracts signals using fourth-order moments. Though various algorithms exist to separate source signals, ICA has two ambiguities: the order of independent components cannot be determined and the sign of independent components is not observed as it does not have any impact in extracting signals.

4.6 Anomaly Detection

Generally, every dataset has data points whose values are within the range specified for that variable. But there exist a few datapoints whose values are out of the range specified and indicates the abnormality. Anomaly detection is a method of distinguishing the data points whose values are beyond the normal range. Those abnormal datapoints are also called *outliers* representing a sign of an opening for a threat. The suspicious data point indicates the possibility of a faulty apparatus, security breach, malfunctioning of sensors, symptoms of disease, etc. In those instances, the anomalous data points need to be removed for further analysis to develop new models.

The types of outliers that are detected using anomaly detection are univariate and multivariate outliers. The univariate outliers are the outliers that occur in a single variable with extreme values deviated from the given range of values for that variable. Multivariate outliers generally occur when the values of multiple variables are combined at one time.

The popular methods under the univariate type are as follows:

> **Standard score (Z-score):** This computes the standard deviation of each attribute of a data point. Then it measures how many values are away from the mean. If a data point with more than three values are away from the mean, it is considered a outlier.

> **Interquartile range:** This is a range between the first quartile (Q1 – 25% of data point are below this quartile) and the third quartile (Q3 – 75% of data point are below this quartile) of a dataset. It is denoted as IQR. If a data point is beyond the range [Q1–1.5 * IQR, Q3 + 1.5 * IQR], then it is called an *outlier*. The constant value by default is 1.5. It can be tuned based on the requirement of application.

Modified Z-scores: It is like a standard Z score with a computing median instead of a standard deviation. This is mainly to determine outliers that are skewed when the mean and standard deviation are used. This method is more robust than others.

The popular methods under the multivariate type are as follows.

In multivariate type, to find the patterns of complex dataset, machine learning algorithms are used.

Local outlier factor (LOF): This type measures the deviation in local density of every sample compared with its neighbors. Outliers are determined based on deviation and which has lower density.

Clustering techniques: In this type, K-means and hierarchical clustering determine the clusters by dividing the dataset into groups. Outliers are identified as the points that are not grouped under any clusters.

Isolation forest: In this type, the isolation trees are determined by dividing the dataset recursively until each instance of data is isolated. The first set of isolated instances are declared as outliers.

Angle-based outlier detection (ABOD): This type works by finding the angle between each point in the dataset. If the angle is different from others, then that point is declared as an outlier.

Based on the application requirement, the usage of the previous types works well.

4.7 Neural Networks

The functionality of the human brain and its interpretation of information is imitated to solve the real-time applications quickly using neural networks. Neural networks consist of processing elements called *nodes* that relate to other nodes through an edge as connection link. This is also called *perceptron*. The weights are assigned to each edge based on the input, and in each iteration, it is updated. After all the data points are fed into the neural networks and the final updation of weights, the network is trained for that specific task.

The neural network architecture consists of layers of nodes that form three different sections: input layer, hidden layer, and output layer. Every node is linked with every other node through connected links. There can be several hidden layers, and the accuracy of the model depends on the number of hidden layers. The weights are associated with every link. There exists a threshold that barricades the activation of nodes. Once the output of a node goes beyond the threshold, the nodes are activated, and data is transferred to the next layer in the network. Until then no data transfer occurs. Neural networks are capable of learning and multitasking, unlike traditional computers that just execute the user instructions sequentially. Neural networks can organize and program themselves with advanced algorithms and derive solutions to unsolved problems yet.

There are various types of neural network models that are applied to real-time applications as follows:

Convolutional neural network (CNN): This is a famous neural network model with one or two convolutional layers based on matrix multiplication to observe and manipulate the patterns in images. These layers work in entirely connected or shared mode. The convolution layer generates feature maps to focus on specific regions of input visuals to investigate deeper and to generate worthful outputs. Hence, they are useful in applications such as natural language processing (NLP), facial recognition, etc.

Deconvolutional neural network: This neural network model is the reverse of CNNs. It is mostly used to find lost signals in the CNN process. It is reverse engineering of the CNN-processed data. It is used to extract features from hierarchical data and mainly used in image synthesis and analysis.

Recurrent neural network: This type of network model has a powerful feature that saves the output and feeds it back to the network node itself to enhance the process and learn from its previous mistakes. Each node acts as a memory cell, which reuses already processed data to recollect the information. It is mainly applied in sales forecasting and text-to-speech applications.

Feed-forward neural network: This type of neural network model is simple, and data flows only in the forward direction from the input layer through various processing nodes until it reaches the output layer node. This is mainly useful for generating clean data from the voluminous and noisy data. It is also called a *multilayer perceptron model*. It is used in computer vision, NLP, etc.

Modular neural network: In this type, several independent neural network modules are operated parallelly with separate inputs to accomplish a portion of an overall objective. The modules do not connect or interfere with one another during computation, but the results are combined to achieve the overall goal.

Generative adversarial network: This type of neural network has two parts, namely, generator and discriminator. The generator generates a new conclusion from the given input, and the discriminator decides the label of the generated conclusion for whether it is real or fake. This network is used to generate realistic sounds, images, text, etc.

4.8 Apriori Algorithm

The apriori algorithm is mainly devised to find the items that are frequently occurring for framing the association rule between items. It is based on prior knowledge of how frequently the items are occurring in the dataset. Association rule mining is how items are related to one another and how an occurrence of an item is influencing the occurrence of another item.

It is given by the if => then format as follows: if people buy bread(A), then there exists more possibility of buying jam(B); this represents single cardinality.

$$A => B$$

If set A contains more than one item, for example A = {x, y, z}, then its cardinality is 3.

There are three measures that are used in apriori algorithms: support, confidence, and lift. *Support* is defined as the ratio of frequency of an item to the total number of occurrences of all items. It can also be calculated for more than one item. For example, if A and B exist, then the frequency of A and B is given by freq(A, B), and the support is given by freq(A,B)/N. The support measure for more than one item is required in market basket analysis. *Confidence* is the ratio of frequency of more than one item to the frequency of a given item, for example freq(A, B)/freq(A). The final measure helps find out how the sale of each item influences the other. *Lift* is defined as the ratio of support measured together with the multiplication of support measured independently, for example, support (A, B)/ (support(A)Xsupport(B)). The apriori algorithm works by using these measures as follows.

1. Find the support and confidence of item sets and select the minimum support and confidence. ·

2. Identify all other supports in the database with the support value higher than the minimum support value.

3. Determine all the association rules of a subset of the item set having a higher confidence than the minimum confidence.

4. Sort the rules in descending order based on its lift value.

5. This algorithm helps in increasing the sales of a store, market basket analysis, etc.

4.9 Singular Value Decomposition (SVD)

The factorization of a matrix A into three matrices to identify the basic patterns is said to be *singular value decomposition*. If the product of a matrix and its transpose gives an identity value, then it is said to be an *orthogonal matrix*. If the matrix has nonzero diagonal values and the rest of the elements are zero, then it is said to be *diagonal matrix*.

Let matrix A be decomposed into three matrices as follows:

$$A = U * S * V^T$$

where U and V are the orthogonal matrices and represent the singular vectors and S is the diagonal matrix that has singular values of A.

The key concept of singular value decomposition is that it splits data into crucial parts and makes use of it to find the patterns in them.

The SVD algorithm is as follows:

1. Eigen decomposition of the matrix $A^T A$ is computed with a standard algorithm.

2. The singular values of A are computed and sorted in descending order.

3. The singular vectors of A are computed for each singular value and normalized to unit length.

The left and right singular vectors of A corresponding to the nonzero singular values are computed with eigen vectors $A A^T$ and $A^T A$, respectively.

4.10 Case Study with Summarization

Here is a case study:

Input file : ChatGPT.txt

```
from google.colab import drive
drive.mount('/content/drive')
```

Mounted at /content/drive
```
import nltk
nltk.download('punkt')
```

[nltk_data] Downloading package punkt to /root/nltk_data...
[nltk_data]   Unzipping tokenizers/punkt.zip.
True
```
import os
os.chdir("/content/drive/MyDrive/Colab Notebooks/LAIoT/
Chapter 4")
print(os.getcwd())
```

/content/drive/MyDrive/Colab Notebooks/LAIoT/Chapter 4

```
file = open("Chapter4.txt", "r")
content = file.read()
print(content)
file.close()
```

ChatGPT (Chat Generative Pre-Trained Transformer) is a chatbot developed by OpenAI and launched on November 30, 2022. Based on a large language model, it enables users to refine and steer a conversation towards a desired length, format, style, level of detail, and language. Successive prompts and replies, known as prompt engineering, are considered at each conversation stage as a context.
By January 2023, it had become what was then the fastest-growing consumer software application in history, gaining over 100 million users and contributing to the growth of OpenAI's valuation to $29 billion. ChatGPT's release spurred the development of competing products, including Bard, Ernie Bot, LLaMA, Claude, and Grok. Microsoft launched its Copilot based on OpenAI's GPT-4. Some observers raised concern about the potential of ChatGPT and similar programs to displace

125

or atrophy human intelligence, enable plagiarism, or fuel misinformation.

ChatGPT is built upon either GPT-3.5 or GPT-4, both of which are members of OpenAI's proprietary series of generative pre-trained transformer (GPT) models, based on the transformer architecture developed by Google and is fine-tuned for conversational applications using a combination of supervised learning and reinforcement learning. ChatGPT was released as a freely available research preview, but due to its popularity, OpenAI now operates the service on a freemium model. It allows users on its free tier to access the GPT-3.5-based version, while the more advanced GPT-4-based version and priority access to newer features are provided to paid subscribers under the commercial name "ChatGPT Plus".

ChatGPT is credited with starting the AI boom, which has led to ongoing rapid and unprecedented development in the field of artificial intelligence.

```
from nltk.tokenize import sent_tokenize
sentence = sent_tokenize(content)
import re
nltk.download('stopwords')  # one time execution
from nltk.corpus import stopwords
```

[nltk_data] Downloading package stopwords to /root/nltk_data...
[nltk_data]   Unzipping corpora/stopwords.zip.

```
corpus = []
for i in range(len(sentence)):
    sen = re.sub('[^a-zA-Z]', " ", sentence[i])
    sen = sen.lower()
    sen = sen.split()
    sen = ' '.join([i for i in sen if i not in stopwords.
```

```
    words('english')])
    corpus.append(sen)
from gensim.models import Word2Vec
all_words = [i.split() for i in corpus]
model = Word2Vec(all_words, min_count=1)
print(all_words)
```

[['chatgpt', 'chat', 'generative', 'pre', 'trained', 'transformer', 'chatbot', 'developed', 'openai', 'launched', 'november'], ['based', 'large', 'language', 'model', 'enables', 'users', 'refine', 'steer', 'conversation', 'towards', 'desired', 'length', 'format', 'style', 'level', 'detail', 'language'], ['successive', 'prompts', 'replies', 'known', 'prompt', 'engineering', 'considered', 'conversation', 'stage', 'context'], ['january', 'become', 'fastest', 'growing', 'consumer', 'software', 'application', 'history', 'gaining', 'million', 'users', 'contributing', 'growth', 'openai', 'valuation', 'billion'], ['chatgpt', 'release', 'spurred', 'development', 'competing', 'products', 'including', 'bard', 'ernie', 'bot', 'llama', 'claude', 'grok'], ['microsoft', 'launched', 'copilot', 'based', 'openai', 'gpt'], ['observers', 'raised', 'concern', 'potential', 'chatgpt', 'similar', 'programs', 'displace', 'atrophy', 'human', 'intelligence', 'enable', 'plagiarism', 'fuel', 'misinformation'], ['chatgpt', 'built', 'upon', 'either', 'gpt', 'gpt', 'members', 'openai', 'proprietary', 'series', 'generative', 'pre', 'trained', 'transformer', 'gpt', 'models', 'based', 'transformer', 'architecture', 'developed', 'google', 'fine', 'tuned', 'conversational', 'applications', 'using', 'combination', 'supervised', 'learning', 'reinforcement', 'learning'], ['chatgpt', 'released', 'freely', 'available', 'research', 'preview', 'due', 'popularity', 'openai', 'operates',

'service', 'freemium', 'model'], ['allows', 'users', 'free',
'tier', 'access', 'gpt', 'based', 'version', 'advanced',
'gpt', 'based', 'version', 'priority', 'access', 'newer',
'features', 'provided', 'paid', 'subscribers', 'commercial',
'name', 'chatgpt', 'plus'], ['chatgpt', 'credited', 'starting',
'ai', 'boom', 'led', 'ongoing', 'rapid', 'unprecedented',
'development', 'field', 'artificial', 'intelligence']]

```
sent_vector=[]
for i in corpus:
    plus=0
    for j in i.split():
        plus+= model.wv[j]
    plus = plus/len(i.split())
    sent_vector.append(plus)
import numpy as np
import pandas as pd
import matplotlib.pyplot as plt
import seaborn as sns
sns.set()
from sklearn.preprocessing import StandardScaler
from sklearn.cluster import KMeans
from sklearn.decomposition import PCA
df = pd.DataFrame(sent_vector)
data_scaler = StandardScaler()
scaled_data = data_scaler.fit_transform(df)
pca = PCA()
pca.fit(scaled_data)
pca.explained_variance_ratio_
```

array([1.81769013e-01, 1.33092284e-01, 1.24260195e-01,

1.15338318e-01, 1.06457539e-01, 8.78785551e-02,
7.66807646e-02, 7.11654797e-02, 5.64657971e-02,
4.68920991e-02, 7.26480201e-16], dtype=float32)

```
plt.figure(figsize=(10,8))
plt.plot(range(1,12),pca.explained_variance_ratio_.cumsum(),
marker='o',linestyle='--')
plt.title("Variance Analysis")
plt.xlabel('Number of sentence vector')
plt.ylabel('Cumulative Explained Variance')
```

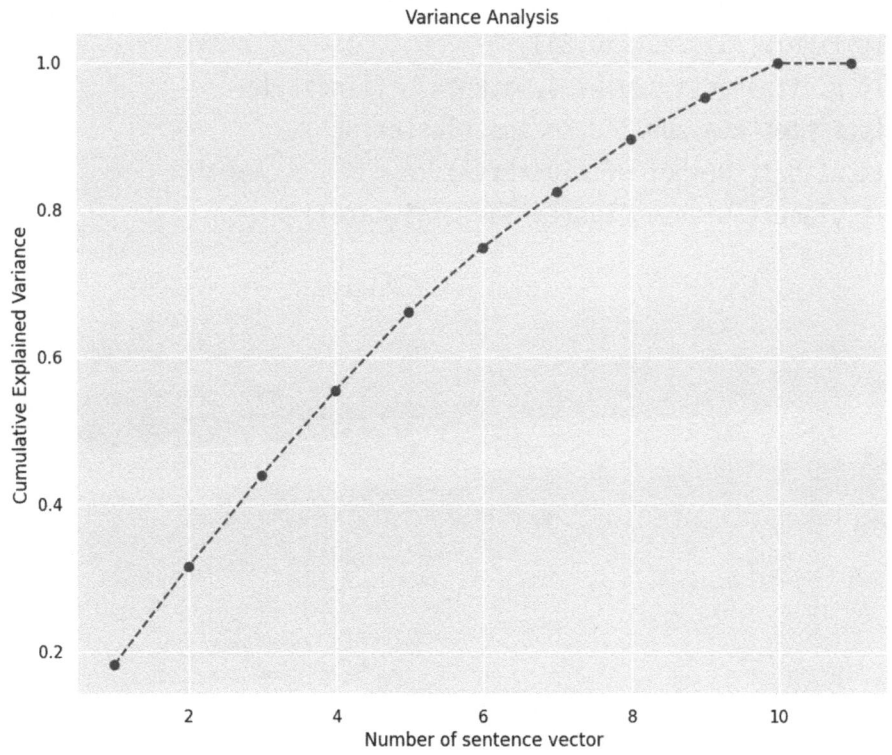

```
pca = PCA(n_components =11)
pca.fit(scaled_data)
PCA(copy=True,iterated_power='auto', n_components=11,
random_state=None, svd_solver='auto',tol=0.0,whiten=False)
scores_pca = pca.transform(scaled_data)
wcss = []
for i in range(1,12):
    kmeans_pca = KMeans(n_clusters=i, init = 'k-means++',
    n_init = 10, random_state = 42)
    kmeans_pca.fit(scores_pca)
    wcss.append(kmeans_pca.inertia_)
plt.figure(figsize=(10,8))
plt.plot(range(1,12),wcss,marker='o',linestyle='--')
plt.title("PCA applied Kmeans clustering")
plt.xlabel('Number of clusters')
plt.ylabel('Within Cluster Sum of Square')
```

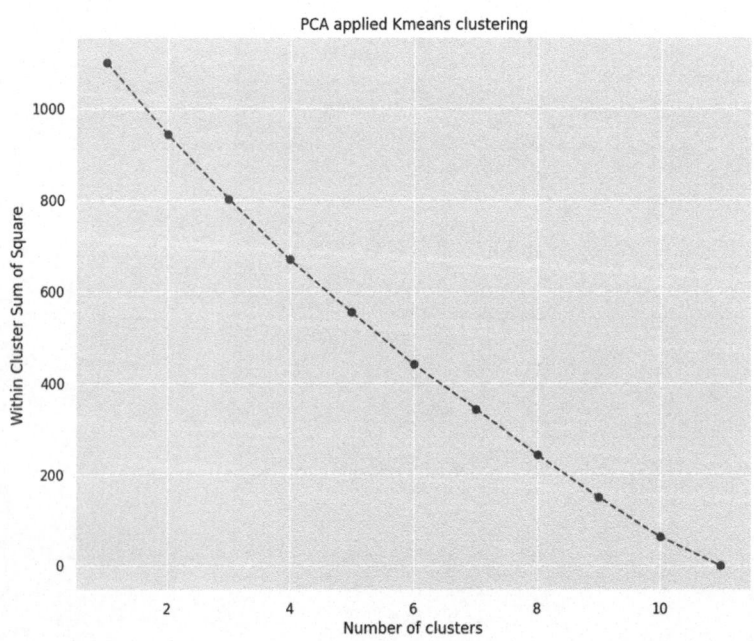

```
n_clusters = 6
kmeans_pca = KMeans(n_clusters, init = 'k-means++',
random_state = 42)
kmeans_pca.fit(scores_pca)
KMeans(algorithm='auto',copy_x=True,init='kmeans++',
max_iter=300,n_clusters=6,n_init=10,random_
state=42,tol=0.0001,verbose=0)
print(df.shape)
print(scores_pca.shape)
print(df.head())
scores_pca_df = pd.DataFrame(scores_pca)
print(scores_pca_df.head())
df_pca_kmeans = pd.concat([df, pd.DataFrame
(scores_pca)],axis=1)
df_pca_kmeans.columns.values[-3:] = [100, 101, 102]
df_pca_kmeans[103] = kmeans_pca.labels_
df_pca_kmeans.rename(columns={100 : "column1" })
df_pca_kmeans.head()
(11, 100)
(11, 10)
         0         1         2         3         4         5         6  \
0 -0.000558 -0.000520 -0.001140 -0.000719  0.004166  0.000213 -0.001123
1  0.000328  0.000260  0.001941  0.000443 -0.001431 -0.001857  0.002054
2  0.002847 -0.001982 -0.002119  0.001163  0.000719 -0.000856  0.001758
3 -0.000510  0.001485 -0.001931 -0.000623 -0.000544 -0.000669 -0.000470
4  0.000483 -0.001095 -0.000232  0.001141 -0.003356 -0.000425  0.001856

         7         8         9 ...        90        91        92        93  \
0  0.004808 -0.002858  0.001510 ...  0.003183  0.001091  0.002733 -0.000012
1  0.001826 -0.002245 -0.002983 ...  0.001318 -0.000291 -0.001805  0.001971
2 -0.000427 -0.001381 -0.000780 ...  0.001636 -0.000502 -0.000776  0.002807
3  0.002278  0.000152 -0.001364 ...  0.001879 -0.000151 -0.000004 -0.001978
4 -0.001289  0.001034 -0.001601 ... -0.001115 -0.000275  0.001000 -0.001199
```

```
            94         95         96         97         98         99
0   0.001408   0.002663   0.001535   0.000791   0.001581  -0.000122
1   0.003436   0.001415  -0.000968   0.000330  -0.000288  -0.002185
2  -0.003008   0.001151   0.000490  -0.001016  -0.001708  -0.001008
3   0.000685  -0.000670  -0.000724  -0.001122   0.000735   0.001186
4   0.001008   0.001520  -0.001825   0.001465  -0.000411  -0.001971

[5 rows x 100 columns]
            0          1          2          3          4          5         6 \
0   0.987164  -5.378297   0.182338   6.673170  -2.845145   3.595698  -2.252162
1  -1.739397   5.983585   0.752204  -1.399546  -1.548659   1.364393  -2.160367
2  -2.669531   5.883429  -3.167400   5.510932  -0.829297  -4.062222  -0.308087
3  -2.627518  -4.332441  -2.156498  -2.917377  -4.536252  -4.219322  -1.615642
4  -4.346540   2.624959   0.069349  -0.489740   0.870991   5.812819   1.896224

            7          8          9
0   1.538730  -0.534066  -1.722997
1   5.781088   2.407620   0.864836
2  -1.987024  -2.434113   0.190779
3  -1.920528   3.576935   0.259598
4  -4.645400   2.295637  -0.576942
0          1          2          3          4          5          6          7
           8          9        ...          1          2          3          4
           5          6        100        101        102        103
0  -0.000558  -0.000520  -0.001140  -0.000719   0.004166   0.000213  -0.001123
   0.004808  -0.002858   0.001510 ...  -5.378297   0.182338   6.673170  -2.845145
   3.595698  -2.252162   1.538730  -0.534066  -1.722997        1
1   0.000328   0.000260   0.001941   0.000443  -0.001431  -0.001857   0.002054
   0.001826  -0.002245  -0.002983 ...   5.983585   0.752204  -1.399546  -1.548659
   1.364393  -2.160367   5.781088   2.407620   0.864836        4
2   0.002847  -0.001982  -0.002119   0.001163   0.000719  -0.000856   0.001758
  -0.000427  -0.001381  -0.000780 ...   5.883429  -3.167400   5.510932  -0.829297
  -4.062222  -0.308087  -1.987024  -2.434113   0.190779        4
```

```
3 -0.000510  0.001485 -0.001931 -0.000623 -0.000544 -0.000669  -0.000470
0.002278    0.000152 -0.001364 ... -4.332441 -2.156498 -2.917377 -4.536252
-4.219322 -1.615642 -1.920528  3.576935  0.259598    2
4  0.000483 -0.001095 -0.000232  0.001141 -0.003356 -0.000425   0.001856
-0.001289  0.001034 -0.001601 ...  2.624959  0.069349 -0.489740 0.870991
5.812819  1.896224 -4.645400  2.295637 -0.576942    4
5 rows × 111 columns
```

```
df_pca_kmeans['test'] = df_pca_kmeans[103].map({0:'first',1:
'second',2:'third',3:'fourth'})
print(df_pca_kmeans.shape)
```

(11, 112)

```
for i in range(100,103):
  x_axis=df_pca_kmeans[i]
  y_axis=df_pca_kmeans[i+1]
  plt.figure(figsize=(10,8))
  sns.scatterplot(x=x_axis,y=y_axis, hue=df_pca_kmeans['test'],
  palette=['g','r','c','m'])
  plt.title('clusters with PCA and Kmeans')
  xlabel = str(i-99)
  ylabel = str(i-98)
  plt.xlabel('PCA attribute'+ xlabel)
  plt.ylabel('PCA attribute'+ ylabel)
  plt.show()
```

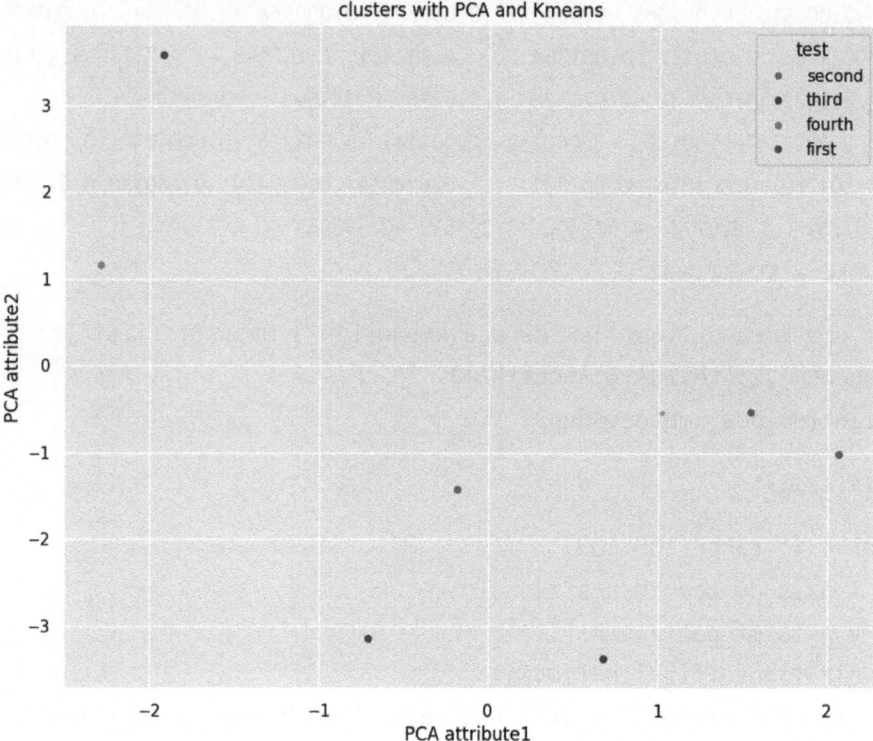

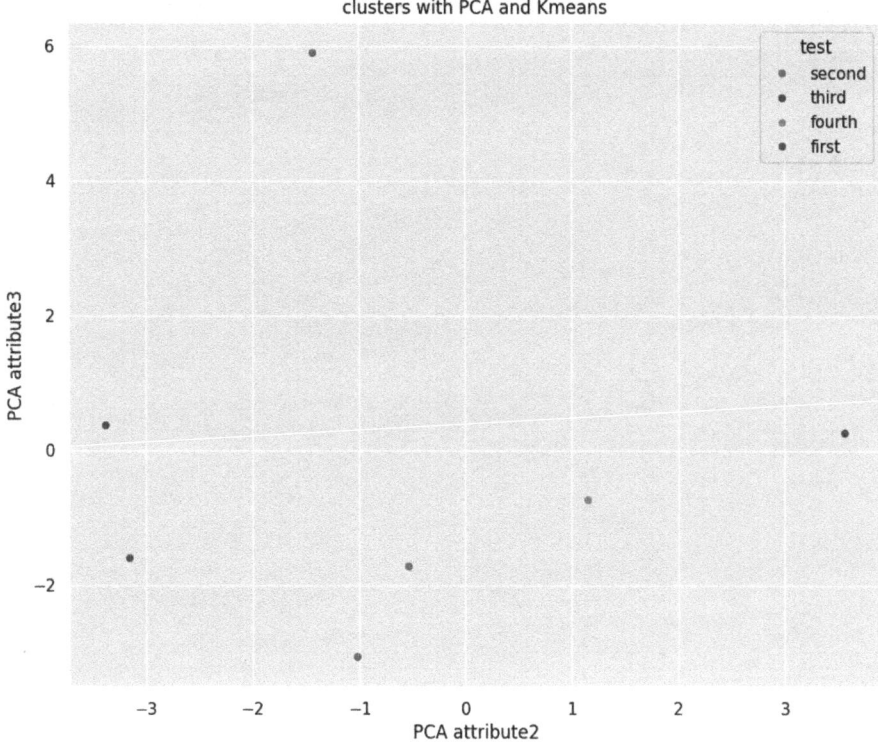

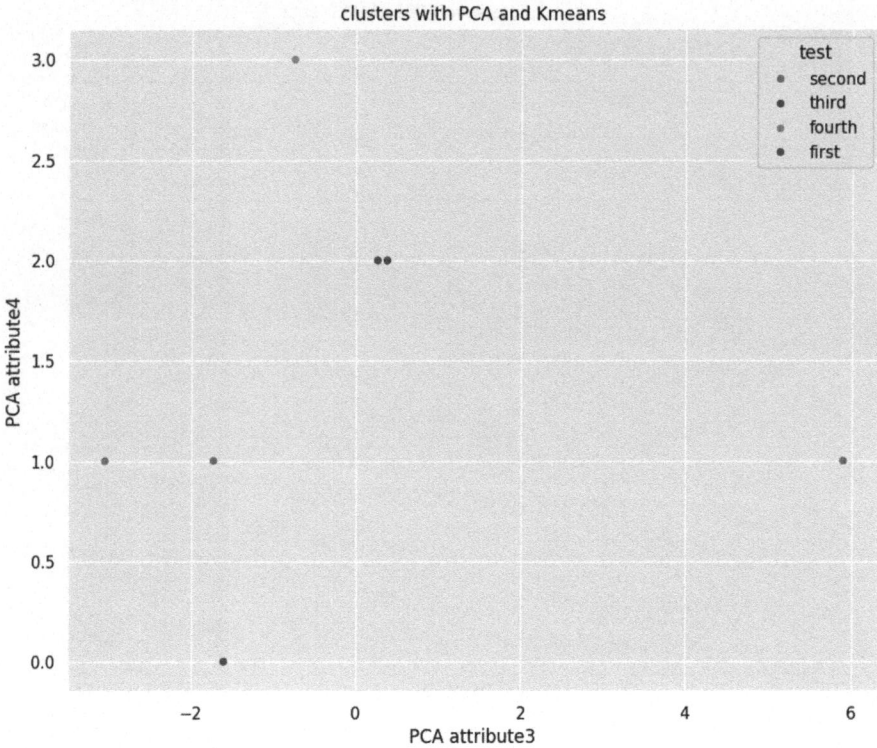

```
from scipy.spatial import distance
my_list=[]
for i in range(n_clusters):
    my_dict={}
    for j in range(len(y_kmeans)):
        if y_kmeans[j]==i:
    my_dict[j] = distance.euclidean(kmeans.cluster_centers_
    [i],sent_vector[j])
    min_distance = min(my_dict.values())
    my_list.append(min(my_dict, key=my_dict.get))
print(my_list)
print(y_kmeans)
```

```
for i in sorted(my_list):
    print(sentence[i])
```

Output:

Some observers raised concern about the potential of ChatGPT and similar programs to displace or atrophy human intelligence, enable plagiarism, or fuel misinformation.

4.11 Summary

This chapter discussed the various unsupervised algorithms and how they work. It explained the clustering techniques to group the data points into different collections. It also discussed PCA, which identifies the main components that are necessary for learning and works for dimensionality reduction, and ICA, which separates the collective data into its independent components. Anomaly detection finds the outliers among the data points. Neural network imitates the human brain to proceed further in the process of learning. The apriori algorithm devises the relationship between the frequently occurring items and their affinity over each other. SVD find the crucial regions in the dataset to figure out the pattern hidden in it.

CHAPTER 5

Reinforcement Learning

5.1 Introduction

The previous two chapters explained the various algorithms of supervised and unsupervised machine learning techniques. There are certain algorithms that imitate the learning process of human beings. Because of the varied nature of these algorithms, instead of categorizing them as supervised or unsupervised, they are categorized under a new machine learning technique called *reinforcement learning*.

The fundamental concept of reinforcement learning is that it must proceed on a trial-and-error basis to achieve the goal. The system determines the reward or penalty based on the actions taken. Ultimately, the goal should be attained with maximum reward points. This approach is highly suitable for real-world problems with complex objectives.

The features of reinforcement learning are as follows:

- The actions of the software part of the system lead to the regulation of the successive output it generates.

- The decision of proceeding further based on actions and output is sequential in nature. Any real number or signal is used to represent the progress and there exists no supervisor to orient towards the goal.

G.R. Kanagachidambaresan and N. Bharathi, *Learning Algorithms for Internet of Things, Maker Innovations Series*, https://doi.org/10.1007/979-8-8688-0530-1_5

5.2 Components of Reinforcement Learning

These are the components of reinforcement learning:

- **Action:** This refers to a decision an agent has made toward the goal from the current state of the environment. The decision is a specific choice from a set of all possible actions from that state to move toward the goal.

- **State:** This represents the current situation of the environment in terms of sensor inputs or variables or features or multi-perspective images.

- **Software agent:** This is responsible for deciding the actions based on the current state and the value of the rewards received from the environment so far.

- **Environment:** This is responsible for deciding the feedback for the agent's actions and for sending the feedback in terms of rewards. Positive feedback is sent if the agent's action led to the goal; otherwise, negative feedback is sent.

- **Reward signal:** The feedback for the agent's actions is represented in terms of numerical values. It depicts how the agent's actions are to the goal.

- **Policy:** This is denoted as μ or π, used by the agent to decide what action needs to be chosen when existing in a state. There are two types of policy based on the nature of the action: the deterministic policy (μ) with the same action for all states and the stochastic policy (π) with different actions from state to another. Also, the policy can be categorized into on-policy and off-policy. On-policy decides the value by itself, and the produced action is performed by the agent. Off-policy decides the value by observing other peers. It has more than one sub policy in which random actions are chosen by behavior policy, and value is evaluated and improved with an estimation policy. Eg. Kids learning from other kids actions to solve their level problems.

- **Value function:** This gives the value of action or state or state-action pair. The value function plays a critical role in deciding how good the action or action-state pair is to progress further toward the goal. There are two types of value function: state value function and action value function for predicting the reward when it exists in any state or performing any action, respectively.

- **Exploration:** This is the process of exploring all possible actions from a state. It is useful in learning about the environment, finding more and more actions, and deciding the better actions to reach the goal faster.

- **Exploitation:** This is the process of choosing the already explored and tried actions to apply the experience learned and to progress toward the goal.

- **Function approximation:** This method addresses the problem of handling very large set of states and actions and complex environments. This method reduces the dimensions of the environment or the state or the actions.

5.3 Types of Reinforcement Learning

Reinforcement learning is categorized into two types: positive and negative reinforcement learning based on the actions and their consequences.

- **Positive reinforcement learning:** If any action or an event followed by an action led to the frequency of the same action happening and consequently increasing the possibility of reaching the goal, then a positive impact is created on the environment. This is known as positive reinforcement learning.

- **Negative reinforcement learning:** If an action or an event followed by an action led to preventing actions or behavior from worsening the possibility of reaching the goal, then a negative impact is avoided on the environment. This is known as negative reinforcement learning.

Further, reinforcement learning is categorized into two types: model-based reinforcement learning and model-free reinforcement learning based on the learning strategy with the environment.

- **Model-based reinforcement learning:** The agent tries to learn the environment behavior, its outcomes, and its rewards. A model to simulate the environment is developed from the learning. The future actions and outcomes can be simulated using the model. Also, the agent can plan and make decisions to estimate the rewards for any actions without even interacting with the environment. The transition probability of states and the outcomes and rewards of each transition are known well in advance before applying them in the real time.

- **Model-free reinforcement learning:** The agent tries to learn using the policy, which guides the agent about the optimal behavior of all possible actions in each state based on a trial-and-error approach. The outcomes and further actions are determined in real time, and they cannot be predicted. In this type, the learning is by observing the consequences of the actions taken and the state of the environment based on the action and not by studying the environment dynamics. Also, this type does not estimate any transition probability of states and its rewards and instead dynamically determines the successive actions and state transitions. This type is quite suitable to the real-world problems for which the solution model is unknown or problems involved in producing accurate outcomes with high complexity.

5.4 Reinforcement Learning Algorithms

The main types of reinforcement learning, as depicted in Figure 5-1, are model free and model based. This chapter focuses on the model-free approach (the model-based approach is covered in previous chapters).

The model-free approach is further categorized into tabular methods and neural network methods. In the tabular method category are dynamic programming, Monte Carlo, temporal difference, and N-step bootstrapping. Deep SARSA, deep Q learning, reinforce and advantage actor critic all take a neural network approach. The following sections will discuss the categories in detail.

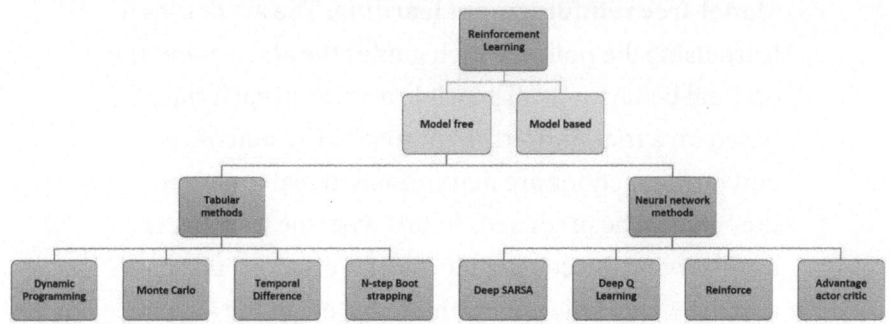

Figure 5-1. *Types of reinforcement learning algorithms*

5.4.1 Markov Decision Process

A mathematical framework to represent the stochastic decision-making system is known as the *Markov decision process* (MDP). MDP is used mainly in real-time problems where the decisions to achieve the goal are random or controlled by a decision-making policy. MDP evaluates the actions taken in the given state of the environment. The effect of any action taken on any state depends only on that state and not on any previous history.

The MDP model comprises the following key components: S is the set of possible states in the environment, A is the set of possible actions, R(s,a) is a reward function that returns a real number when action *a* is taken from the state *s*, and T is a note on the effect of each action on each state.

5.4.2 Bellman Equation

The value of the state when using the policy π is given by the expected value returned when in state *s* at time instant t:

$$v_{\pi}(s) = E\ [G_t \mid S_t = s]$$

where S_t is the set of possible states and G_t is the return. It can be expanded as follows:

$$= E\ [R_{t+1} + \gamma R_{t+2} + ... + \gamma^{T-t-1}R_T\ |\ S_t = s]$$
$$= E\ [R_{t+1\ +}\ \gamma G_{t+1}\ |\ S_t = s]$$
$$= \sum_a \pi(a|s) \sum_{s',r} p(s,a)[r + \gamma v_\pi(s')]$$

where R or r is the reward, γ is the discount, p is the probability, π is the policy, a is the action, and s/s' are the states.

Similarly, the value of the action when using the policy π is as follows:

$$q_\pi(s,a) = E\ [G_t\ |\ S_t = s,\ A_t = a]$$

$$= E\ [R_{t+1} + \gamma R_{t+2} + ... + \gamma^{T-t-1}R_T\ |\ S_t = s,\ A_t = a]$$
$$= E\ [R_{t+1\ +}\ \gamma G_{t+1}\ |\ S_t = s,\ A_t = a]$$
$$= \sum_{s',r} p(s,a)\left[r + \gamma \sum_{a'} \pi(s')q_\pi(s',a')\right]]$$

where A_t is the set of actions.

5.4.3 Tabular Methods

Methods that solves less complex problems with arrays or tables as states and actions along with any policy to generate value table by the agent are called *tabular methods*. Dynamic programming, Monte Carlo, temporal difference, and N-step bootstrapping are tabular methods.

5.4.4 Dynamic Programming

The policy that is controlling the actions from the states is used in dynamic programming (DP) to find the optimal values using value functions. The policy and value are interdependent, and hence the optimal policy or value can be determined if any exists. There are two approaches for finding the optimal value or policy: value iteration and policy iteration. Both types of algorithms are using the bellman equations as their rules to update values for the improvement of the policy or value. The limitations of the DP are the unavailability of model for most real-time problems and the high computational cost.

- **Value iteration:** This is a process of determining the optimized policy by iterating the value estimation with a random initial value. Parallelly executing all possible actions to decide the maximum value of action makes the computation heavier than policy iteration.

$$\pi_*(s) = \sum_{s',r} p(s,a)\left[r + \gamma v_*(s')\right]$$

The value update rule is as follows:

$$V(s) \leftarrow \sum_{s',r} p(s,a)\left[r + \gamma V(s')\right]$$

where r is reward, γ is discount, p is the probability, π is the policy, a is the action, s/s' are the states, and v is the value of the state.

- **Policy Iteration:** This is a process of alternatively improving the policy and the estimated values.

    Iterative policy evaluation is as follows:

$$V(s) \leftarrow \sum_{s',r} p(s,a)\left[r + \gamma V(s')\right]$$

After each iteration, the Δ is computed as the difference between the current value and the previous value of the state under the given policy.

$$\Delta \leftarrow \max\left(\Delta, |v\text{-}V(s)|\right)$$

As it proceeds with successive iterations, it improves closer to the optimal policy $v_\pi(s)$, and the policy improvement is as follows:

$$v_\pi(s) = \sum_a \pi(a|s) \sum_{s',r} p(s,a)\left[r + \gamma v_\pi(s')\right]$$

Based on the different actions taken from the same state, the policy differs from π to π′, and the policy improvement for the different actions is as follows:

$$\text{If } q_\pi(s, \pi'(s)) \geq v_\pi(s) \text{ then } v_{\pi'}(s) \geq v_\pi(s)$$

5.4.5 Monte Carlo Methods

The Monte Carlo (MC) methods are based on computing the estimates of action values of any policy, and the action values are used to search the new better policy iteratively. The final step is the yield of optimal policy that generates optimized actions to achieve the goal. As MC is categorized as model-free reinforcement learning, the optimal policy is determined by the experience gained and not from any models. The prediction in this method refers to the value of state and action to find the higher value and in turn optimal policy. A simple and common way to estimate the values is by averaging the return value collected after visiting each state. The analogy behind this is that if we visit a state more often, then the average return value will be nearer to the value of that state.

On-policy Monte Carlo method: This method estimates the value based on the policy used for control. The behavior policy and the target policy are the same in this method.

Here the value update rule is as follows:

$$Q(s,a) \leftarrow \text{average } (G(S_t, A_t)$$

Off-policy Monte Carlo method: This method estimates the value based on all possible actions with exploration using the random policy, and learned experiences are used to optimize the policy used for control. The behavior policy and the target policy are different.

Here the value update rule is as follows:

$$Q(s,a) \leftarrow Q(s,a) + \frac{w_t}{C(s,a)}\left[G - Q(s,a)\right]$$

where $C(s,a) = \sum_{k=1}^{N} w_k$ and w_t is the importance sampling.

5.4.6 Temporal Difference

The reinforcement learning technique predicts the expected value over the sequence of states by combining dynamic programming and the Monte Carlo method. This technique ensures the convergence to the optimal value with probability one and is one of the best reinforcement learning methods. it works in a way such that updating the state value with an immediate reward R_{t+1} and the reward prediction $V(S_{t+1})$ at the next instance of time.

$$V(S_t) \leftarrow V(S_t) + \alpha\,[R_{t+1} + \gamma V(S_{t+1}) - V(S_t)]$$

where $R_{t+1} + \gamma V(S_{t+1}) - V(S_t)$ is Temporal Difference Error denoted as δ_t

SARSA: This is an on-policy temporal difference control that learns the action value function instead of the state value function. In other words, the learning is focused on the transition from one state action pair to another state action pair and its action values. The quintuple is given by $(S_t, A_t, R_{t+1}, S_{t+1}, A_{t+1})$ and is repeated as it progresses.

The updating of action value is as follows:

$$Q\,(S_t, A_t) \leftarrow Q\,(S_t, A_t) + \alpha\,[R_{t+1} + \gamma\,Q\,(S_{t+1}, A_{t+1}) - Q\,(S_t, A_t)]$$

where α is the learning rate and γ is the discount factor.

- **Q-learning:** This is an off-policy temporal difference control that uses two policies: target policy and exploratory policy. The target policy is based on ϵ, a greedy policy that selects the action with a maximum value for state transitions. The exploratory policy is used to explore all possible actions, and their outcomes are learned as experiences. The update rule of action value is as follows:

$$Q\,(S_t,\, A_t) \leftarrow Q\,(S_t,\, A_t) + \alpha\,[R_{t+1} + \gamma\,Q(S_{t+1}, a) - Q\,(S_t,\, A_t)]$$

5.4.7 N-Step Bootstrapping

The Monte Carlo and temporal difference methods was discussed in the previous two sections. N-step bootstrapping is the technique used in the family of methods between the Monte Carlo and temporal difference methods. SARSA is the method where n=1 and the update rule are based on only the next transition or step as mentioned. If n is equal to or greater than the actual duration of the episode (n episode is the set of state transitions from initial state to the final state), then it is Monte Carlo method as mentioned in its update rule. So, this technique is used by the family of methods that fall under the n= (2, 3 .... n-1) step bootstrapping method. The challenge in this method is to choose the optimized value for n based on the environment of real-world problems to avoid bias and variance. If we choose closer to n=1, then the learning is good in terms of control and determining the estimates of values but not in the right direction toward the goal. On the other hand, if the n is closer to the episode, then the variance between the value estimates is high but moves toward the right direction toward the goal.

N-step SARSA: This is an extended approach of SARSA on the policy temporal difference method. In this method, the update rule is as follows:

$$Q\,(S_t,\, A_t) \leftarrow Q\,(S_t,\, A_t) + \alpha\,[\,G_{t:t+n} - Q\,(S_t,\, A_t)]$$

where $G_{t:t+n} = R_{t+1} + \gamma R_{t+2} + ... + \gamma^n Q(S_{t+1}, A_{t+1})$.

5.4.8 Neural Network Methods

An artificial neural network is a system that imitates the learning mechanism of the human brain. A neural network is a collection of neurons, connected in an organized way across layers. It comprises the input layer, more than one hidden layer, and finally an output layer. There are basically two types of neural network: feed forward neural network and feed backward network.

The general concept of using neural network and updating q values takes place in two steps. First, the experience learned by the agent by interacting with the environment is collected in the replay memory as a transition tuple (S, A, R, S'). The experience collected is then used to redefine the neural network parameters θ to improve the performance of the network. Second, the improvement in the network is carried out by finding the estimated mean squared error (loss function) using the estimated q values and the actual values. The neural network parameters are updated based on the current value of it and the loss function value.

Deep SARSA: This method is the combination of SARSA and neural network. In this method, the table is not used to store the estimates of action values (Q). Alternatively, it is generated by the neural network with the set of all possible states as input. The output of the neural network is a vector that has estimates of Q values for all the actions from a particular state.

The neural network optimization is as follows:

Loss function $L(\theta) = \dfrac{1}{|K|} \sum_{i=1}^{|k|} \left[R_i + \gamma \hat{q}(\theta_{targ}) - \hat{q}(\theta) \right]^2$

The neural network parameter θ is updated using stochastic gradient descent as follows:

$$\theta = \theta - \alpha \nabla L(\theta)$$

Deep Q-learning: This method is the combination of Q-Learning and neural network. As mentioned in the Q-learning, two policies are involved in determining the actions: target policy and exploratory policy. But the Q values are estimated by the neural network, and it is not stored in the Q-table. The neural network generates the estimates of Q values for all the states and actions.

Here is the loss function: $L(\theta) = \frac{1}{|K|}\sum_{i=1}^{|k|}\left[R_i + \gamma\hat{q}\left(\theta_{targ}\right) - \hat{q}(\theta)\right]^2$

The neural network parameter θ is updated like Deep SARSA as $\theta = \theta - \alpha \nabla L(\theta)$.

REINFORCE: This algorithm is the combination of policy gradient method and Monte Carlo method along with neural network. Here the neural network parameter θ update rule is like Deep SARSA and Q-learning and is given by $\theta = \theta - \alpha \nabla J(\theta)$.
where $\nabla J(\theta)$ is the gradient of the performance estimate of the policy.

$$\nabla J(\theta) \propto \sum_{s}\mu(s)\sum_{a}q_\pi(s,a)\nabla\pi(a|,s|,\theta)$$

and according to the Monte Carlo method, the gradient is as follows:
$\nabla J(\theta) = \gamma^t G_t \frac{\nabla\pi\left(A_t|,S_t|,\theta_t\right)}{\pi\left(A_t|,S_t|,\theta_t\right)}$ and the update rule is as follows:

$$\theta_{t+1} = \theta_t + \alpha\gamma^t G_t\nabla lnln\,\pi\left(A_t|S_t,\ \theta_t\right)$$

where $\nabla lnln\,\pi\left(A_t|S_t,\ \theta_t\right) = \frac{\nabla\pi\left(A_t|,S_t|,\theta_t\right)}{\pi\left(A_t|,S_t|,\theta_t\right)}$

Advantage Actor-Critic: This algorithm is the combination of policy gradient method and temporal difference method along with neural network. The advantage function is given as follows:

$$\mathrm{Adv}_\pi(s,a) = q_\pi(s,a) - v_\pi(s)$$

Alternatively, it can be written as $Adv(s,a) = r(s,a) + \gamma v(s') - v(s)$ in terms of reward and discount. The update rule is as follows:

$\theta_{t+1} = \theta_t + \alpha \gamma^t Adv_t \nabla \ln\ln \pi(A_t | S_t, \theta_t)$ if $Adv_t > 0$, then taking the action is better than following the policy.

5.5 Applications of Reinforcement Learning

There are several applications that are suitable for applying reinforcement learning. Any problem that is more complex to train using any model and supervised or unsupervised can be resolved by reinforcement learning.

- **Autonomous cars:** The dynamic nature of traffic, speed, road conditions, etc., are key factors that the car system should learn from its experiences with the principles of exploration and exploitation. It needs to consider several factors for the decision-making process.

- **Computerized medical diagnosis:** Deep reinforcement learning uses the learning framework to learn the process of generating medical reports, diagnosing the diseases with symptoms and reports, etc. It learns with the features of deep neural networks and from the sequence of actions, which is maximizing the reward to progress through whatever desired.

- **Adaptive traffic light control:** The adaptive traffic control is very much required for the increase in the number of cars in cities. Here the decision-making involves the dynamic rate of arrival of cars and the traffic on directions, which might change dynamically. Reinforcement learning is a better approach to adaptively control the traffic lights by learning from the dynamic environment.

- **Personalized recommendations for customers:**
  The reinforcement learning approach is completely
  different from the traditional way of making
  recommendations to customers.

- **Gaming:** Reinforcement learning is monopolizing
  the gaming domain. When compared to traditional
  gaming solutions such as complex behavioral tree,
  reinforcement learning is simpler for implementing
  the logic of the game. The system can learn easily by
  simulating the environment and observe the outcomes
  of all possible actions to reach the goal of the game.

5.6 Case Study with Python

This demonstration uses a built-in library called gym that has various
reference environments. The reference environment used in this program
is Frozen Lake version 1 under the environment category Toy Text. The
main API functions of the environment are reset(), step(), render(), and
close(). The key attributes are action space and observation space (states).
The basic steps for creating the Frozen Lake environment is taken from
the gym library and modified according to the requirements for this
demonstration.

The Frozen Lake environment is the one in which the lake is denoted
by map with 4x4 or 8x8 squares as states. The possible state values are
frozen, hole, start, and goal. The player moves through the states until
the goal is reached or falls in a hole. The possible actions allowed are left,
down, right, and up.

#The following are the libraries used in the program.

```
import gym
import matplotlib
import matplotlib.pyplot as plt      # for visualizing the
frozen lake environment
from matplotlib import animation      # for display video
from IPython.display import HTML      # for display video
import os
from typing import Tuple, Dict, Optional, Iterable,
Callable    # in render Optional
```

#The image files required to display the environment are in the folder in the google drive.
#The following code connects the runtime environment of google colab to google drive folder.

```
from google.colab import drive
drive.mount('/content/drive')
os.chdir("drive/My Drive/Colab Notebooks/")
print(os.getcwd())
print(os.path.dirname(os.path.realpath('__file__')))
# @title FrozenLakeEnv Class - Run this cell by pressing
"Shift + Enter"
from contextlib import closing
from io import StringIO
from os import path
from typing import List, Optional
import numpy as np
from gym import Env, logger, spaces, utils
from gym.envs.toy_text.utils import categorical_sample
from gym.error import DependencyNotInstalled
import os
```

```
LEFT = 0
DOWN = 1
RIGHT = 2
UP = 3
MAPS = {
    "4x4": ["SFFF", "FHFH", "FFFH", "HFFG"],
    "8x8": [
        "SFFFFFFF",
        "FFFFFFFF",
        "FFFHFFFF",
        "FFFFFHFF",
        "FFFHFFFF",
        "FHHFFFHF",
        "FHFFHFHF",
        "FFFHFFFG",
    ],
}
# DFS to check that it's a valid path.
```

#This function checks whether the path it chooses to progress is reaching the goal or hole.

```
def is_valid(board: List[List[str]], max_size: int) -> bool:
    frontier, discovered = [], set()
    frontier.append((0, 0))
    while frontier:
        r, c = frontier.pop()
        if not (r, c) in discovered:
            discovered.add((r, c))
            directions = [(1, 0), (0, 1), (-1, 0), (0, -1)]
            for x, y in directions:
                r_new = r + x
```

```
                c_new = c + y
                if r_new < 0 or r_new >= max_size or c_new < 0
                or c_new >= max_size:
                    continue
                if board[r_new][c_new] == "G":
                    return True
                if board[r_new][c_new] != "H":
                    frontier.append((r_new, c_new))
    return False
```

The below code generates the random map for frozen lake problem with one start and one goal along with holes and frozen regions.

```
def generate_random_map(size: int = 8, p: float = 0.8) ->
List[str]:

    """Generates a random valid map (one that has a path from
    start to goal)
    Args:
        size: size of each side of the grid
        p: probability that a tile is frozen
    Returns:
        A random valid map
    """

    valid = False
    board = []  # initialize to make pyright happy
    while not valid:
        p = min(1, p)
        board = np.random.choice(["F", "H"], (size, size),
        p=[p, 1 - p])
        board[0][0] = "S"
        board[-1][-1] = "G"
```

```
        valid = is_valid(board, size)
    return ["".join(x) for x in board]
```

Frozen Lake class under gym environment. We can add code blocks in the key functions of this class

```
class FrozenLakeEnv(gym.Env):
    """
```

Frozen lake involves crossing a frozen lake from Start(S) to Goal(G) without falling into any Holes(H) by walking over the Frozen(F) lake.

The agent may not always move in the intended direction due to the slippery nature of the frozen lake.

Action Space

The agent takes a 1-element vector for actions.

The action space is `(dir)`, where `dir` decides direction to move in which can be:

- 0: LEFT

- 1: DOWN

- 2: RIGHT

- 3: UP

Observation Space

The observation is a value representing the agent's current position as

*current_row * nrows + current_col (where both the row and col start at 0).*

*For example, the goal position in the 4x4 map can be calculated as follows: 3 * 4 + 3 = 15.*

The number of possible observations is dependent on the size of the map.

For example, the 4x4 map has 16 possible observations.

Rewards

Reward schedule:
- Reach goal(G): +1
- Reach hole(H): 0
- Reach frozen(F): 0
Arguments
` ` `

*gym.make('FrozenLake-v1', desc=None, map_name="4x4", is_*
slippery=True)
` ` `

`desc`: Used to specify custom map for frozen lake. For
example,
    desc=["SFFF", "FHFH", "FFFH", "HFFG"].
    A random generated map can be specified by calling the
    function `generate_random_map`. For example,
    ` ` `

    *from gym.envs.toy_text.frozen_lake import generate_*
    random_map
    *gym.make('FrozenLake-v1', desc=generate_random_*
    map(size=8))
    ` ` `

`map_name`: ID to use any of the preloaded maps.
    "4x4":[
        "SFFF",
        "FHFH",
        "FFFH",
        "HFFG"
        ]
    `is_slippery`: True/False. If True will move in
    intended direction with
probability of 1/3 else will move in either perpendicular
direction with

```
equal probability of 1/3 in both directions.
    For example, if action is left and is_slippery is
    True, then:
    - P(move left)=1/3
    - P(move up)=1/3
    - P(move down)=1/3
### Version History
* v1: Bug fixes to rewards
* v0: Initial versions release (1.0.0)
"""

# Rendering modes
metadata = {
    "render_modes": ["human", "ansi", "rgb_array"],
    "render_fps": 4,
}

# initialization of key parameters

def __init__(
    self,
    render_mode: Optional[str] = None,
    desc=None,
    map_name="4x4",
    is_slippery=True,
    shaped_rewards: bool = False,
    size: int = 16
):
    if desc is None and map_name is None:
        desc = generate_random_map()
    elif desc is None:
        desc = MAPS[map_name]
    self.desc = desc = np.asarray(desc, dtype="c")
```

```
self.nrow, self.ncol = nrow, ncol = desc.shape
self.reward_range = (0, 1)
self.goal = 15
self.maze = self._create_Lake(size=size)
self.shaped_rewards = shaped_rewards
self.distances = np.full((16), np.inf)
nA = 4
nS = nrow * ncol
__file__ = os.path.dirname(os.path.realpath
('__file__'))
self.initial_state_distrib = np.array(desc == b"S").
astype("float64").ravel()
self.initial_state_distrib /= self.initial_state_
distrib.sum()
self.P = {s: {a: [] for a in range(nA)} for s in
range(nS)}

# moves to that state
def to_s(row, col):
    return row * ncol + col

# increment of the state - goes to next state.
def inc(row, col, a):
    if a == LEFT:
        col = max(col - 1, 0)
    elif a == DOWN:
        row = min(row + 1, nrow - 1)
    elif a == RIGHT:
        col = min(col + 1, ncol - 1)
    elif a == UP:
        row = max(row - 1, 0)
    return (row, col)
```

updation of the probability matrix based on the actions taken and the closeness to the goal.

```
def update_probability_matrix(row, col, action):
        newrow, newcol = inc(row, col, action)
        newstate = to_s(newrow, newcol)
        newletter = desc[newrow, newcol]
        terminated = bytes(newletter) in b"GH"
        reward = float(newletter == b"G")
        return newstate, reward, terminated
    for row in range(nrow):
        for col in range(ncol):
            s = to_s(row, col)
            for a in range(4):
                li = self.P[s][a]
                letter = desc[row, col]
                if letter in b"GH":
                    li.append((1.0, s, 0, True))
                else:
                    if is_slippery:
                        for b in [(a - 1) % 4, a, (a +
                        1) % 4]:
                            li.append(
                                (1.0 / 3.0, *update_
                                probability_matrix
                                (row, col, b))
                            )
                    else:
                        li.append((1.0, *update_
                        probability_matrix(row, col, a)))
    self.observation_space = spaces.Discrete(nS)
    self.action_space = spaces.Discrete(nA)
```

```
        self.render_mode = render_mode
        # pygame utils
        self.window_size = (min(64 * ncol, 512), min
        (64 * nrow, 512))
        self.cell_size = (
            self.window_size[0] // self.ncol,
            self.window_size[1] // self.nrow,
        )
        self.window_surface = None
        self.clock = None
        self.hole_img = None
        self.cracked_hole_img = None
        self.ice_img = None
        self.elf_images = None
        self.goal_img = None
        self.start_img = None
```

```
# This function returns the tuple after an is taken. The tuple
contains state, reward, whether it reaches goal or hole,
whether no.of episodes reached, probability.
```

```
    def step(self, a)-> Tuple[Tuple[int], float, bool, Dict]:
        #print(type(self.P[self.s][a]))
        transitions = self.P[self.s][a]
        #print(type(transitions))
        i = categorical_sample([t[0] for t in transitions],
        self.np_random)
        p, s, r, t = transitions[i]
        self.s = s
        self.lastaction = a
        if self.render_mode == "human":
            self.render()
        return (int(s), r, t, False, {"prob": p})
```

This function resets the state as the start state.

```
def reset(
    self,
    *,
    seed: Optional[int] = None,
    options: Optional[dict] = None,
):
    super().reset(seed=seed)
    self.s = categorical_sample(self.initial_state_distrib,
    self.np_random)
    self.lastaction = None
    if self.render_mode == "human":
        self.render()
    return int(self.s), {"prob": 1}
```

This function is used to to call render_gui method of frozen lake environment

```
def render(self):
    if self.render_mode is None:
        logger.warn(
            "You are calling render method without
            specifying any render mode. "
            "You can specify the render_mode at
            initialization, "
            f'e.g. gym("{self.spec.id}", render_mode="rgb_
            array")'
        )
    elif self.render_mode == "ansi":
        return self._render_text()
    else:  # self.render_mode in {"human", "rgb_array"}:
        return self._render_gui(self.render_mode)
```

This function is used to generate visualization of frozen lake environment

```python
    def _render_gui(self, mode):
        try:
            import pygame
        except ImportError:
            raise DependencyNotInstalled(
                "pygame is not installed, run `pip install
                gym[toy_text]`"
            )
        if self.window_surface is None:
            pygame.init()
            if mode == "human":
                pygame.display.init()
                pygame.display.set_caption("Frozen Lake")
                self.window_surface = pygame.display.set_
                mode(self.window_size)
            elif mode == "rgb_array":
                self.window_surface = pygame.Surface(self.
                window_size)
        assert (
            self.window_surface is not None
        ), "Something went wrong with pygame. This should never
        happen."
        if self.clock is None:
            self.clock = pygame.time.Clock()
        if self.hole_img is None:
            file_name = os.path.join(os.getcwd(), "img/
            hole.png")
            self.hole_img = pygame.transform.scale(
                pygame.image.load(file_name), self.cell_size
            )
```

```
if self.cracked_hole_img is None:
    file_name = os.path.join(os.getcwd(), "img/cracked_
    hole.png")
    self.cracked_hole_img = pygame.transform.scale(
        pygame.image.load(file_name), self.cell_size
    )
if self.ice_img is None:
    file_name = os.path.join(os.getcwd(), "img/
    ice.png")
    self.ice_img = pygame.transform.scale(
        pygame.image.load(file_name), self.cell_size
    )
if self.goal_img is None:
    file_name = os.path.join(os.getcwd(), "img/
    goal.png")
    self.goal_img = pygame.transform.scale(
        pygame.image.load(file_name), self.cell_size
    )
if self.start_img is None:
    file_name = os.path.join(os.getcwd(), "img/
    stool.png")
    self.start_img = pygame.transform.scale(
        pygame.image.load(file_name), self.cell_size
    )
if self.elf_images is None:
    elfs = [
        os.path.join(os.getcwd(), "img/elf_left.png"),
        os.path.join(os.getcwd(), "img/elf_down.png"),
        os.path.join(os.getcwd(), "img/elf_right.png"),
        os.path.join(os.getcwd(), "img/elf_up.png"),
    ]
```

```
    self.elf_images = [
        pygame.transform.scale(pygame.image.load
        (f_name), self.cell_size)
        for f_name in elfs
    ]
desc = self.desc.tolist()
assert isinstance(desc, list), f"desc should be a list
or an array, got {desc}"
for y in range(self.nrow):
    for x in range(self.ncol):
        pos = (x * self.cell_size[0], y * self.cell_
        size[1])
        rect = (*pos, *self.cell_size)
        self.window_surface.blit(self.ice_img, pos)
        if desc[y][x] == b"H":
            self.window_surface.blit(self.hole_
            img, pos)
        elif desc[y][x] == b"G":
            self.window_surface.blit(self.goal_
            img, pos)
        elif desc[y][x] == b"S":
            self.window_surface.blit(self.start_
            img, pos)
        pygame.draw.rect(self.window_surface,
        (180, 200, 230), rect, 1)
# paint the elf
bot_row, bot_col = self.s // self.ncol, self.s %
self.ncol
cell_rect = (bot_col * self.cell_size[0], bot_row *
self.cell_size[1])
```

```
    last_action = self.lastaction if self.lastaction is not
    None else 1
    elf_img = self.elf_images[last_action]
    if desc[bot_row][bot_col] == b"H":
        self.window_surface.blit(self.cracked_hole_img,
        cell_rect)
    else:
        self.window_surface.blit(elf_img, cell_rect)
    if mode == "human":
        pygame.event.pump()
        pygame.display.update()
        self.clock.tick(self.metadata["render_fps"])
    elif mode == "rgb_array":
        return np.transpose(
            np.array(pygame.surfarray.pixels3d
            (self.window_surface)), axes=(1, 0, 2)
        )
```

```
# This function simulates the step function based on reward

    def simulate_step(self, state: Tuple[int], action: int):
        reward = self.compute_reward(state, action)
        next_state = self._get_next_state(state, action)
        done = next_state == self.goal
        info = {}
        return next_state, reward, done, info
```

```
#Reward is calculated based on the action taken

    def compute_reward(self, state: Tuple[int], action: int)
    -> float:
        next_state = self._get_next_state(state, action)
        if self.shaped_rewards:
```

```
        return - (self.distances[next_state] / self.
        distances.max())
    return - float(state != self.goal)
def _get_next_state(self, state: Tuple[int], action: int)
-> Tuple[int]:
    if action == 0:
        if(state != 0) and (state != 4) and (state != 8)
        and (state != 12):
            next_state = state -1
        else: next_state = state
    elif action == 1:
        if(state != 12) and (state != 13) and
        (state != 14):
            next_state = state+4
        else: next_state = state
    elif action == 2:
        if(state != 3) and (state != 7) and (state != 11):
            next_state = state+1
        else: next_state = state
    elif action == 3:
        if(state != 0) and (state != 1) and (state != 2)
        and (state != 3):
            next_state = state-4
        else: next_state = state
    else:
        raise ValueError("Action value not
        supported:", action)
    if next_state in self.observation_space:
        return next_state
    return state

# The environment is created by passing size as parameter
```

```
@staticmethod
def _create_Lake(size: int) -> Dict[Tuple[int],
Iterable[Tuple[int]]]:
  lake = {s: [] for s in range(size) }
  return lake
```

This function computes the distances from the all other states to the goal

```
@staticmethod
def _compute_distances(goal: Tuple[int],
                       lake: Dict[Tuple[int],
                       Iterable[Tuple[int]]]) ->
                       np.ndarray:
    distances = np.full((16), np.inf)
    visited = set()
    distances[goal] = 0.
    while visited != set(lake):
        sorted_dst = [(v // 4, v % 4) for v in distances.
        argsort(axis=None)]
        closest = next(x for x in sorted_dst if x not in
        visited)
        visited.add(closest)
        for neighbour in lake[closest]:
            distances[neighbour] =
            min(distances[neighbour],
            distances[closest] + 1)
    return distances
```

This function computes the dimension for small rectangle in he center for visualization of environment

```
@staticmethod
def _center_small_rect(big_rect, small_dims):
    offset_w = (big_rect[2] - small_dims[0]) / 2
    offset_h = (big_rect[3] - small_dims[1]) / 2
    return (
        big_rect[0] + offset_w,
        big_rect[1] + offset_h,
    )
```

This is the function to display text in the visualization.
The text can be probability of the action from a state.

```
def _render_text(self):
    desc = self.desc.tolist()
    outfile = StringIO()
    row, col = self.s // self.ncol, self.s % self.ncol
    desc = [[c.decode("utf-8") for c in line] for line
    in desc]
    desc[row][col] = utils.colorize(desc[row][col], "red",
    highlight=True)
    if self.lastaction is not None:
        outfile.write(f"  ({['Left', 'Down', 'Right', 'Up']
        [self.lastaction]})\n")
    else:
        outfile.write("\n")
    outfile.write("\n".join("".join(line) for line in
    desc) + "\n")
    with closing(outfile):
        return outfile.getvalue()
```

This function will be called when the visualization window
is closed.

```
    def close(self):
        if self.window_surface is not None:
            import pygame
            pygame.display.quit()
            pygame.quit()
```

```
#Main code to execute the reinforcement learning using policy
iteration.
```

```
env = FrozenLakeEnv(render_mode='rgb_array')
env.reset()
#print("this is frozen lake")
#env.render()
frame = env.render()
plt.figure(figsize=(6,6))
plt.axis('off')
plt.imshow(np.squeeze(frame))
print(f"For example, the initial state is: {env.reset()}")
print(f"The space state is of type: {env.observation_space}")
print(f"An example of a valid action is: {env.action_space.
sample()}")
print(f"The action state is of type: {env.action_space}")
```

```
env.reset()
action = env.action_space.sample()
print(action)
next_state, reward, done, info,_ = env.step(action)
frame = env.render()
plt.axis('off')
plt.title(f"State: {next_state}")
plt.imshow(np.squeeze(frame))
```

```
state = env.reset()
episode = []
terminated = False
truncated = False

while not terminated:
    action = env.action_space.sample()
    next_state, reward, terminated, truncated, _ = env.step
    (action)
    episode.append([state, action, reward, terminated,
    truncated,next_state])
    state = next_state
env.close()
print(f"Congrats! You just generated your first episode:\
n{episode}")

policy_probs = np.full((16, 4), 0.25)
print(policy_probs)

# policy function

def policy(state):
    action_probabilities = np.array([0.25, 0.25, 0.25, 0.25])
    return action_probabilities

action_probabilities = policy((0,0))
for action, prob in zip(range(4), action_probabilities):
    print(f"Probability of taking action {action}: {prob}")

state = env.reset()
action_probabilities = policy(state)
objects = ('Up', 'Right', 'Down', 'Left')
y_pos = np.arange(len(objects))
plt.bar(y_pos, action_probabilities, alpha=0.5)
```

```
plt.xticks(y_pos, objects)
plt.ylabel('P(a|s)')
plt.title('Random Policy')
plt.tight_layout()
plt.show()

# Policy evaluation, policy improvement and policy
iteration methods

state_values = np.zeros(shape=(16))
def policy_evaluation(policy_probs, state_values, theta=1e-6,
gamma=0.99):
    delta = float("inf")
    while delta > theta:
        delta = 0
        for index in range(16):
            old_value = state_values[(index)]
            new_value = 0
            action_probabilities = policy_probs[(index)]
            for action, prob in enumerate(action_
            probabilities):
                next_state, reward, _, _ = env.simulate_
                step((index), action)
                new_value += prob * (reward + gamma *
                state_values[next_state])
            state_values[(index)] = new_value
            delta = max(delta, abs(old_value - new_value))

def policy_improvement(policy_probs, state_values, gamma=0.99):
    policy_stable = True
    for index in range(16):
        old_action = policy_probs[(index)].argmax()
        new_action = None
```

```python
            max_qsa = float("-inf")
            for action in range(4):
                next_state, reward, _, _ = env.simulate_step
                ((index), action)
                qsa = reward + gamma * state_values[next_state]
                if qsa > max_qsa:
                    max_qsa = qsa
                    new_action = action
            action_probs = np.zeros(4)
            action_probs[new_action] = 1.
            policy_probs[(index)] = action_probs
            if new_action != old_action:
                policy_stable = False
    return policy_stable

def policy_iteration(policy_probs, state_values, theta=1e-6,
gamma=0.99):
    policy_stable = False
    while not policy_stable:
        policy_evaluation(policy_probs, state_values,
        theta, gamma)
        policy_stable = policy_improvement(policy_probs,
        state_values, gamma)

# calling of policy iteration method

policy_iteration(policy_probs, state_values)

# Displaying the actions and state transitions as video

def display_video(frames):
    orig_backend = matplotlib.get_backend()
    matplotlib.use('Agg')
    fig, ax = plt.subplots(1, 1, figsize=(5, 5))
```

```
matplotlib.use(orig_backend)
ax.set_axis_off()
ax.set_aspect('equal')
ax.set_position([0, 0, 1, 1])
im = ax.imshow(frames[0])
def update(frame):
    im.set_data(frame)
    return [im]
anim = animation.FuncAnimation(fig=fig, func=update,
frames=frames,
                                    interval=50, blit=True,
                                    repeat=False)
return HTML(anim.to_html5_video())
```

The agent method for testing the environment and play game with successive actions for 10 episodes by default.

```
def test_agent(environment, policy, episodes=10):
    frames = []
    for episode in range(episodes):
        state = env.reset()
        done = False
        frames.append(env.render())
        while not done :
            p = policy(state)
            if isinstance(p, np.ndarray):
                action = np.random.choice(4, p=p)
            else:
                action = p
            #action = np.argmax(p)
            next_state, reward, done, extra_info,_ = env.step
            (action)
            img = env.render()
```

```
            frames.append(img)
            state = next_state
        if state==15:
          break;
    print(episode)
    return display_video(frames)

# calling the agent method to test the enviroment
state = env.reset()
test_agent(env, policy, episodes=100)
```

5.7 Summary

Reinforcement learning is the process of gaining experience much like a
human being does. The components involved are state, action, reward,
return, exploration, exploitation, experience, etc. Two major types of
reinforcement learning are the model-based and model-free methods. We
discussed model-free methods in this chapter. Also, the applications of RL
and a case study with Python are discussed. In the next chapter, we will
discuss the artificial neural networks in more detail.

CHAPTER 6

Artificial Neural Networks for IoT

6.1 Introduction to Artificial Neural Networks (ANNs)

An ANN is a deep learning method that imitates the neural functionality of the human brain. The human brain is composed of billions of neurons that consist of dendrites, axon, and nucleus. Information is communicated between the excited neurons through electric and chemical signals. Figure 6-1 shows the synapse or synaptic junction formed through the connection of the axon of one neuron and the dendrites of another neuron. The axon and dendrites acts as the neuro transmitter and receiver during information transmission in the human brain. The information received through dendrites is processed at the nucleus and in turn produces electric and chemical signals to transmit to the next neuron through axon tips when its threshold level is reached.

© G.R. Kanagachidambaresan and N. Bharathi 2024
G.R. Kanagachidambaresan and N. Bharathi, *Learning Algorithms for Internet of Things,*
Maker Innovations Series, https://doi.org/10.1007/979-8-8688-0530-1_6

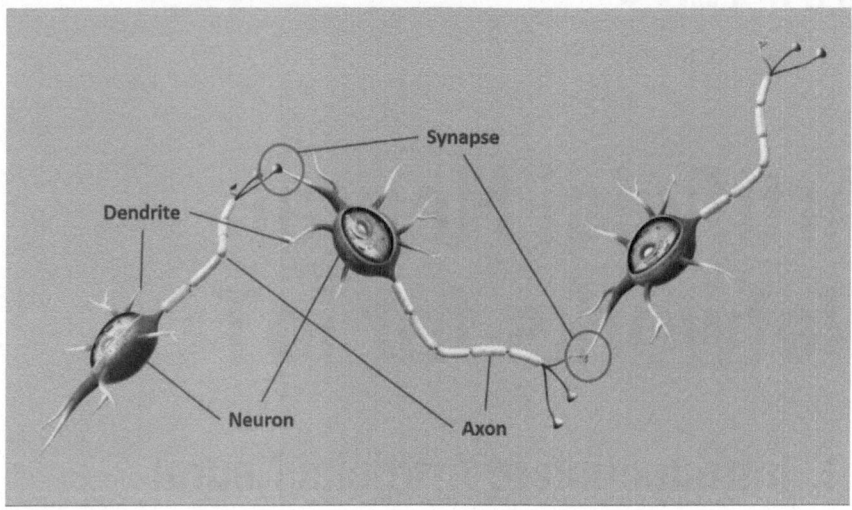

Figure 6-1. *Communication structure of the human brain*

An artificial neural network is similar to the structure of neurons in a human brain. But an ANN works well with structured data. An ANN comprises interconnected nodes as neurons connected through several layers. The weights over the connection between neurons are updated during the learning process. When it reaches the threshold level, the weighted sum of all inputs are transmitted to the next level of nodes, and this process continues until it reaches the output layer. At the output layer, activation functions are applied on the weighted sum to map the decision or prediction. The connection strengths are shown through the weights, and the prediction is based on the activation function. A perceptron in one such node forms the neural network, as shown in Figure 6-2. In that figure, x1, x2...xn are inputs; w1, w2...w3 are weights; and b is bias, which controls the behavior of the network by determining the instance of triggering the activation function. Finally, the activation function is applied to the neuron output, which is the weighted sum of all inputs and the bias. Many perceptron connected with each other form the complex ANN, which can make predictions, recognize patterns, and do other tasks in machine learning and deep learning.

We have discussed the similarity of human brains and ANNs; however, though the functionality is similar, there are differences. The major difference is that the human brain does the processing parallelly, but the ANN does the processing in a series, with lots of layers based on the requirements. Among the layers, one does the processing in microseconds (slower), and the next one does the computation in nanoseconds (faster).

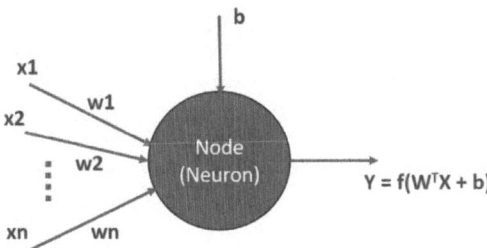

Figure 6-2. *Perceptron*

6.2 Architecture of ANN

An ANN has three components: three different layers, weights over the connections, and the bias and activation function. Among the three layers, the input layer is the first layer, which consists of input neurons that collect input from the environment and pass it to the hidden layer through connections. The connections have weights from each input neuron to each hidden layer neurons. The hidden neurons compute the sum of the multiplied values of the previous layer input and its weight and the bias. The computation is known as *transformation*, which can then be passed to the output layer through the activation function. There can be more than one hidden layer. The two features of the hidden layer help in learning more complex computations such as the depth and width of the hidden layer. The depth represents the number of sublayers in the hidden layer, and the width depicts the number of neurons in each hidden sublayer. The ANN is classified into five different types, covered next.

6.3 Activation Function

The activation function is responsible for triggering the neurons to calculate the output. The triggering is decided when the sum of weighted inputs and bias reach the threshold and are passed through the activation function. The output is calculated based on the triggered neurons and passed to the next layer until it reaches the output layer. It is called *forward propagation*. The output is compared with the actual output, and the error is calculated. The weights are updated based on the error calculated, which is called *back propagation*. This process continues for a number of iterations (*epochs*) to reduce errors as well as improve the prediction of the output.

The various activation function types are (i) linear functions, (ii) binary step functions, and (iii) nonlinear function. A *linear activation function* is simple and proportional to the input. As it has no major impact on the weighted sum and bias, it is also called an *identity function* or *no activation*. It is denoted mathematically as $f(x) = x$ or $y = x$. The pitfall is that there is no meaning for the back propagation or the increase in number of layers in the neural network with the linear activation function as there is no change in the output.

The *binary step activation function* is based on the threshold value, which is the deciding value to activate the neuron. If the input value at the neuron is greater than the threshold, then it is activated and passed to the next hidden layer. If the input value is less than the threshold, then it is deactivated. It is denoted mathematically as $f(x) = 0$ for $x < 0$ and $f(x) = 1$ for $x \geq 0$. The constraint in this function is that it is not suitable for multiclass problems, and also since the gradient is zero, it is not very suitable for back propagation.

The *nonlinear activation functions* are better than enabling models to generate output from input through complex computations unlike simple linear functions. The traditional nonlinear functions are sigmoid and hyperbolic tangent. The *sigmoid* function is used when probability is predicted as an output. It is denoted mathematically as $f(x) = 1/(1+e^{-x})$. The $f(x)$ value ranges between 0 and 1. The *hyperbolic tangent* function is used when the $f(x)$ value ranges from -1 to 1, and it is denoted mathematically as $f(x) = (e^x - e^{-x}) / (e^x + e^{-x})$. It is also called the *tanh function*. The output of this function is always zero centered, and it is easier to map positive, negative, and neutral values.

The modern nonlinear activation functions are Rectified Linear Unit (ReLU), Leaky ReLU, Exponential LU (ELU), Softmax, and Swish. *ReLU* allows for back propagation although it seems to be linear. It does not activate all the neurons at once; instead, neurons whose linear transformation is less than zero are deactivated. It is denoted mathematically as $f(x) = \max(0,x)$. The downside of ReLU is the Dying ReLU problem, which creates dead neurons due to the zero gradient value at the negative side. A few neurons or none at all update their values during back propagation. *Leaky ReLU* is mathematically denoted as $f(x) = \max(0.1x,x)$. It is an enhanced version of ReLU that solves the dying ReLU problem. It has small slope in the negative region, enabling the update of a few neuron values during back propagation and hence prevents the creation of dead neurons.

Exponential LU is also an enhanced version of ReLU, which uses the log curve for negative values. It is denoted mathematically as $f(x) = x$ for $x \geq 0$ and $f(x) = \alpha(e^x-1)$ for $x < 1$. It avoids dead neurons and helps the network weights and biases in the right direction toward the goal. The *softmax* function is the combination of more than one sigmoid. It is denoted mathematically as $f(x_i)=\exp(x_i)/(\sum_j \exp(x_j))$. It computes the relative probabilities and proceeds with the probability of each class. It is more suitable for multiclass classification as an activation function at the last layer. The *swish* function is more suitable for deep neural networks for challenging domains such as image classification, object classification, etc. It is denoted mathematically as $\sigma(x) = x/(1+ e^{-x})$ or $f(x) = x * \text{sigmoid}(x)$. It is bounded in the negative region and unbounded in the positive region. It was developed by experts at Google.

The following Python code demonstrates the implementation of the activation functions discussed in this chapter, and Figure 6-3 shows their output.

```python
import matplotlib.pyplot as plt
import numpy as np

x = [-5,-4.5,-4,-3.5,-3,-2.5,-2,-1.5,-1,-0.5,0,1,1.5,2,2.5,3,
3.5,4,4.5,5]
y1,y2,y3,y4,y5,y6,y7,y8,y9 = [],[],[],[],[],[],[],[],[]

def linear(x):
    return x

def binary(x):
    if x >= 0:
      return 1
    else:
      return 0

def sigmoid(x):
    return 1 / (1 + np.exp(-x))

def tanh(x):
    return np.tanh(x)

def relu(x):
    return max(0.0, x)

def leakyrelu(x):
    alpha = 0.1  #small negative slope
    return max(alpha*x, x)

def exponentialLU(x):
    alpha = 1.0 # positive value hyper parameter
    if x > 0:
```

```
        return x
    else:
      return (1 * (np.exp(x) - 1))

# for softmax activation function to find sum of exponential
of inputs
def exp_sum(x):
    exp_x = 0
    for I in range(len(x)):
      exp_x += np.exp(x[i])
    return exp_x

def softmax(x,exp_sm):
    exp_x = np.exp(x)
    return exp_x / exp_sm

def swish(x):
    return x * sigmoid(x)

exp_sm = exp_sum(x)

for i in range(len(x)):
    y1.append(linear(x[i]))
    y2.append(binary(x[i]))
    y3.append(sigmoid(x[i]))
    y4.append(tanh(x[i]))
    y5.append(relu(x[i]))
    y6.append(leakyrelu(x[i]))
    y7.append(exponentialLU(x[i]))
    y8.append(softmax(x[i],exp_sm))
    y9.append(swish(x[i]))

plt.plot(x,y1, label='Linear')
plt.plot(x,y2, label='Binary')
plt.legend()
```

```
plt.show()
plt.plot(x,y3, label='Sigmoid')
plt.plot(x,y4, label='Tanh')
plt.legend()
plt.show()
plt.plot(x,y5, label='ReLU')
plt.plot(x,y6, label='Leaky ReLU')
plt.plot(x,y7, label='Exponential LU')
plt.legend()
plt.show()
plt.plot(x,y8, label='Softmax')
plt.plot(x,y9, label='Swish')
plt.legend()
plt.show()
```

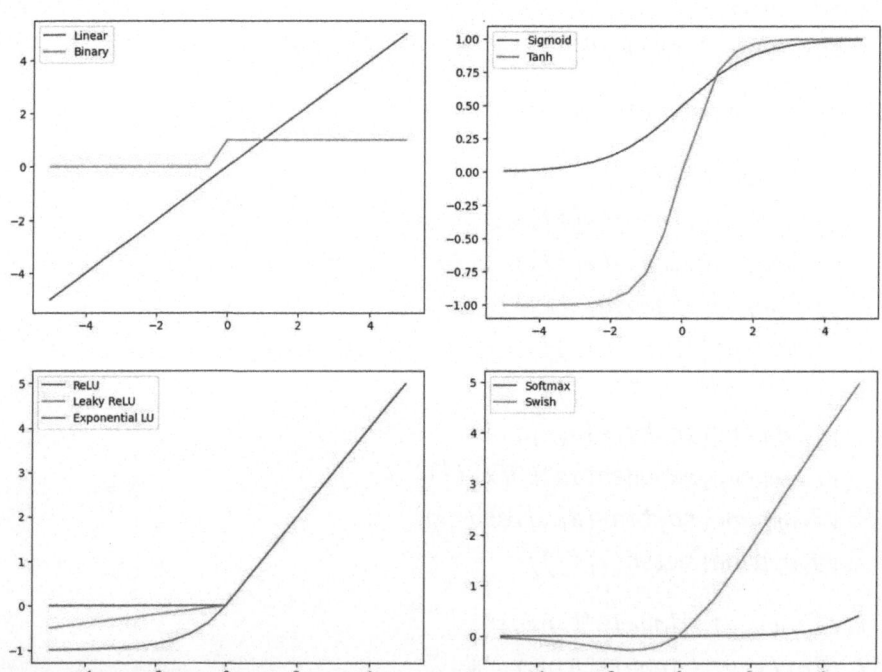

Figure 6-3. *Implementation of various activation functions*

6.4 Loss Function

The loss function represents the deviation of the predicted output value from the actual output. Its value is the measure for evaluating the model's fit for the given dataset. If the value is higher, the predictions are not good. If the value is lower, the predictions are good as the loss value, and good predictions are inversely proportional. The improvement or worsening of the model is depicted by the loss value across the epochs. The various types of loss functions are as follows:

Regression loss function: The loss function used in regression models is called the *regression loss function*. Generally, the regression type of models is to predict continuous numerical values. They are well suitable for deducing the relationship between more than one independent variable and a dependent variable.

The regression loss functions are as follows:

- **Mean Squared Error Loss:** This denotes the magnitude of the error between the predicted value through ML models and actual value. Mathematically it is denoted as **MSE = $(\sum(Y_{pred} - Y_{actual})^2)$ / n**. This loss is also known as L2 loss.

- **Mean Squared Logarithmic Error Loss:** This is defined as the average of the difference between the log value of actual and predicted outputs. It is denoted as MSLE = $(\sum(\log(Y_{actual} + 1) - \log(Y_{pred} + 1))^2)$ / n

- **Mean Absolute Error Loss:** This loss function treats all errors equally irrespective of their magnitude. It computes average of the absolute difference between the predicted value and actual value as **MAE = $(\sum|Y_{pred} - Y_{actual}|)$ / n**. It is also called *L1 loss*.

Binary class loss function: The loss function used in binary class models is called the *binary class loss function*. In this type, the predicting output is one among the two values, namely, 0 or 1. The threshold value is decided to classify the data under the category 0 or 1. The default threshold value will be 0.5, and the prediction under category 0 if the value is less than the threshold or under category if value is greater than threshold.

The binary classification loss functions is as follows:

- **Binary Cross-Entropy Loss:** This is a performance measure of binary classification models. Entropy is the degree of disorder in the system. It is denoted by $BCE = (\sum - (y_i * \log(p_i) + (1-y_i) * \log(1-p_i)))/ n$. p_i is the probability of class 1 and $(1-p_i)$ is the probability of class 0.

- **Hinge Loss:** This is used in support vector machines (SVM) algorithms. It is defined as the amount of error in prediction increases exponentially if the prediction is wrong for the far away points from the margin. It is denoted by $HL = \max(0, 1-y_{pred}*y_{actual})$.

- **Squared Hinge Loss:** This is the same as hinge loss but squares the value to make it more sensitive to identify outliers and useful in gradient based optimization as it is differentiable.

Multiclass loss function: This loss function is used in multiclass classification models. Normally, in the multiclass models, the number of target variables or classes are more than two. Multiple threshold values will be used to categorize the data based on the predicted values. It is the extended model of binary class in which more than two categories exists.

The multiclass classification loss functions are as follows:

- **Multiclass cross-entropy loss:** This is defined as the summation of values that formed by multiplying the actual value and the logarithm of the predicted value. It is mathematically denoted as $L(y,p) = -\sum(y_i * \log(p_i))/n$. It is also known as softmax loss or categorical cross-entropy loss.

- **Sparse multiclass cross-entropy loss:** This is like multiclass cross-entropy loss, but the usage of it is different. If the target labels can be represented as one hot-encoded vector, then multiclass cross entropy is suitable, but it is not feasible at certain times when there are more target classes involved in which sparse multiclass cross entropy is used. The constraint of this sparse cross entropy is that the target classes are mutually exclusive and not like [0.3,0.7,0.1] soft probabilities for three classes. One sample belongs to only one class when multiclasses are involved.

- **Kullback-Leibler divergence loss:** This is an error measure of the relative difference between the two probability distributions. It is denoted as $KL(P\|Q) = \sum P(x)\log(P(x)) - P(x)\log(Q(x))$.

The following Python code demonstrates the implementation of the regression and binary class loss functions discussed.

```
import numpy as np
y_actual = np.array([1.0, 0.0, 1.0, 0.0, 1.0])
y_pred = np.array([0.8, 0.1, 0.9, 0.3, 0.9])
# Calculates the mean squared error (MSE) loss between
predicted and actual values.
def mseloss(y_actual, y_pred):
```

```
    n = len(y_actual)
    mse_loss = np.sum((y_pred - y_actual) ** 2) / n
    return mse_loss

# Calculates the mean squared logarithmic error (MSLE) loss.
def msleloss(y_actual, y_pred):
    n = len(y_actual)
    msle_loss = np.sum((np.log(y_actual+1) - np.log(y_pred+1))
    ** 2) / n
    return msle_loss

# Calculates the mean absolute error (MAE) loss
def maeloss(y_actual, y_pred):
    mae_loss = np.mean(np.abs(y_pred - y_actual))
    return mae_loss

# Calculates the binary cross-entropy loss
def bceloss(y_actual, y_pred):
    n = len(y_actual)
    bce_loss= -np.sum(y_actual * np.log(y_pred)+(1 - y_actual)
    *np.log(1 - y_pred)) / n
    #print(bce_loss)
    return bce_loss

# Calculates hinge loss
def hingeloss(y_actual, y_pred):
    # replacing 0 = -1
    new_predicted = y_pred - (y_pred == 0)
    new_actual = y_actual - (y_actual == 0)
    hinge_loss = np.mean([max(0, 1-x*y) for x, y in zip
    (new_actual, new_predicted)])
    return hinge_loss

#Calculates the squared hinge loss
```

```
def sqhingeloss(y_actual, y_pred):
    return hingeloss(y_actual, y_pred) * hingeloss
    (y_actual, y_pred)

#Displaying the loss calculated
print(mseloss(y_actual, y_pred))
print(msleloss(y_actual, y_pred))
print(maeloss(y_actual, y_pred))
print(bceloss(y_actual, y_pred))
print(hingeloss(y_actual, y_pred))
print(sqhingeloss(y_actual, y_pred))
```

Output:

```
0.031999999999999994
0.01885637600036785
0.15999999999999998
0.1791800084452842
0.56
0.31360000000000005
```

The following Python code demonstrates the implementation of multiclass loss functions discussed earlier:

```
import numpy as np

# calculate the multi class cross entropy loss
def multiclassCELoss(y_actual, y_pred):
    mcce_loss = -1/len(y_actual) * np.sum(np.sum(y_actual *
    np.log(y_pred)))
    return mcce_loss

# calculate sparse multi class cross entropy loss
def sparse_categorical_crossentropy(y_actual, y_pred):
    # convert true labels to one-hot encoding
```

```
    y_actual_onehot = np.zeros_like(y_pred)
    y_actual_onehot[np.arange(len(y_actual)), y_actual] = 1
    # calculate loss
    scce_loss = -np.mean(np.sum(y_actual_onehot * np.log
    (y_pred), axis=-1))
    return scce_loss
```

```
# calculate Kullback Leibler divergence Loss
def kl_divergence_loss(y_actual, y_pred):
    return np.sum(y_actual * np.log(y_actual/y_pred, where=y_
    actual > 0), where = y_actual  > 0)
```

```
# define true labels and predicted probabilities as
NumPy arrays
y_actual = np.array([[0.0, 1.0, 0.0], [0.0, 0.0, 1.0],
[1.0, 0.0, 0.0]])
y_pred = np.array([[0.1, 0.6, 0.3], [0.2, 0.3, 0.5],
[0.8, 0.1, 0.1]])
```

```
print(multiclassCELoss(y_actual, y_pred))
```

```
y_actual = np.array([1,2,0])
y_pred = np.array([[0.1, 0.6, 0.3], [0.2, 0.3, 0.5],
[0.8, 0.1, 0.1]])
```

```
print(sparse_categorical_crossentropy(y_actual, y_pred))
```

```
y_actual = np.array([1.0, 0.0, 1.0, 0.0, 1.0])
y_pred = np.array([0.8, 0.1, 0.9, 0.3, 0.9])
```

```
print(kl_divergence_loss(y_actual, y_pred))
```

   output:

0.47570545188004854
0.47570545188004854
0.43386458262986244

6.5 Types of Artificial Neural Network Architectures

The interconnection between the neurons is essential for orchestrating the processing elements. The ways the neurons are arranged and their interconnections highly influence the computation time. Based on the interconnection architecture, the ANN is categorized into five types.

1. Single-layer feed forward network

2. Multilayer feed-forward network

3. Single node with its own feedback

4. Single-layer recurrent network

5. Multilayer recurrent network

6.5.1 Feed-Forward ANN

The first two types (single-layer and multilayer feed forward networks) are categorized under feed forward ANNs because the interconnections are only in the forward direction. There is no feedback interconnection from the output layer to the hidden or output layer.

Single-layer feed-forward network

The single layer feed*forward network consists of two layers: input and output. The input does not perform any computation and only passes the input to the network. The connections connecting the input and output layers carries weights and are involved in the computation of output. The output layer receives weights along with input values, and based on the activation of each neuron in this layer, it computes the output value.

The following code illustrates the single-layer feed-forward network with two input neurons and one output neuron.

```python
import numpy as np

# Define the input data
X = np.array([[0, 0], [0, 1], [1, 0], [1, 1]])

# Define the output data
y = np.array([0, 1, 1, 1])

# Define the weights
w = np.zeros(2)     # as X has 2 features

# Define the bias, learning rate and number of epochs
b = 0
lr = 0.1
epochs = 100

# Train the model
for epoch in range(epochs):
  for i in range(len(X)):

    # Calculate the output
    output = np.dot(X[i], w) + b

    # Calculate the error
    error = y[i] - output

    # Update the weights and bias
    w += lr * error * X[i]
    b += lr * error

# Print the final weights and bias
print("Weights:", w)
print("Bias:", b)
```

```
# Predict the output for the input data
for i in range(len(X)):
  output = np.dot(X[i], w) + b
  print("Input:", X[i], "Output:", output)
```

Here is the output:

```
Weights: [0.44426104 0.47202838]
Bias: 0.2780323894561509
Input: [0 0] Output: 0.2780323894561509
Input: [0 1] Output: 0.7500607655160025
Input: [1 0] Output: 0.7222934263956347
Input: [1 1] Output: 1.1943218024554862
```

The following code illustrates this with three input neurons in the input layer and two neurons in output layer.

```
import numpy as np

# Define the activation function
def sigmoid(x):
  return 1 / (1 + np.exp(-x))

# Define the derivative of the activation function
def d_sigmoid(x):
  return sigmoid(x) * (1 - sigmoid(x))

# Initialize the weights and biases
weights = np.array([[0.1, 0.2],[0.3,0.4], [0.5, 0.6]])
biases = np.array([0.1, 0.2])

# Define the input data
X = np.array([[0.1, 0.2, 0.3], [0.4, 0.5, 0.6]])

# Calculate the output of the first layer
layer_1 = sigmoid(np.dot(X, weights) + biases)
```

```
# Define the second layer weights and biases
weights_2 = np.array([[0.7, 0.8], [0.9, 1.0]])
biases_2 = np.array([0.3, 0.4])

# Calculate the output of the second layer
layer_2 = sigmoid(np.dot(layer_1, weights_2) + biases_2)

# Print the output of the second layer
print(layer_2)
```

Here's the output:

```
[[0.77928423 0.81475486]
 [0.79882622 0.83384928]]
```

Multilayer feed-forward network

Multilayer feed forward networks have more than two layers. In addition to the input and output layers, the hidden layer is also included. The back propagation algorithm is also used to adjust weights and improve the accuracy. The following code illustrates this with two input neurons in the input layer, three neurons in the single hidden layer, and one neuron in the output layer. The number of hidden layers can be extended further if more data exists and high accuracy is required.

```
import numpy as np
import matplotlib.pyplot as plt
import matplotlib.colors
from sklearn.model_selection import train_test_split
from sklearn.metrics import accuracy_score, mean_squared_error
from sklearn.preprocessing import OneHotEncoder
from sklearn.datasets import make_blobs
from tqdm import tqdm_notebook
#creation of color map
my_cmap = matplotlib.colors.LinearSegmentedColormap.
from_list("", ["red","blue","green","yellow"])
```

```python
#Generating 500 data with 4 labels
data, labels = make_blobs(n_samples=500, centers=4,
n_features=2, random_state=0)
print(data.shape, labels.shape)

#Display the data using scatterplot
plt.scatter(data[:,0], data[:,1], c=labels, cmap=my_cmap)
plt.show()

#Conversion of the 4 labels - multi-class to 2 labels binary
labels_orig = labels
labels = np.mod(labels_orig, 2)
plt.scatter(data[:,0], data[:,1], c=labels, cmap=my_cmap)
plt.show()

#Preparation of the binary data for training
X_train, X_val, Y_train, Y_val = train_test_split(data, labels,
stratify=labels, random_state=0)
print(X_train.shape, X_val.shape)
class simpleFFNetwork:

  #intialize the parameters
  def __init__(self):
    self.wt1 = np.random.randn()
    self.wt2 = np.random.randn()
    self.wt3 = np.random.randn()
    self.wt4 = np.random.randn()
    self.wt5 = np.random.randn()
    self.wt6 = np.random.randn()
    self.wt7 = np.random.randn()
    self.wt8 = np.random.randn()
    self.wt9 = np.random.randn()
    self.bs1 = 0
```

```
    self.bs2 = 0
    self.bs3 = 0
    self.bs4 = 0

  def sigmoid(self, x):
    return 1.0/(1.0 + np.exp(-x))

  def forward_pass(self, x):
    #forward pass - preactivation and activation
    self.x1, self.x2 = x
    self.a1 = self.wt1*self.x1 + self.wt2*self.x2 + self.bs1
    self.hn1 = self.sigmoid(self.a1)
    self.a2 = self.wt3*self.x1 + self.wt4*self.x2 + self.bs2
    self.hn2 = self.sigmoid(self.a2)
    self.a3 = self.wt5*self.x1 + self.wt6*self.x2 + self.bs3
    self.hn3 = self.sigmoid(self.a3)
    self.a4 = self.wt7*self.hn1 + self.wt8*self.hn2 + self.
    wt9*self.hn3 + self.bs4
    self.hn4 = self.sigmoid(self.a4)
    return self.hn4

  def grad(self, x, y):
    #back propagation
    self.forward_pass(x)

    self.dwt7 = (self.hn4-y) * self.hn4*(1-self.hn4) * self.hn1
    self.dwt8 = (self.hn4-y) * self.hn4*(1-self.hn4) * self.hn2
    self.dwt9 = (self.hn4-y) * self.hn4*(1-self.hn4) * self.hn3
    self.dbs4 = (self.hn4-y) * self.hn4*(1-self.hn4)

    self.dwt1 = (self.hn4-y) * self.hn4*(1-self.hn4) * self.wt7
    * self.hn1*(1-self.hn1) * self.x1
    self.dwt2 = (self.hn4-y) * self.hn4*(1-self.hn4) * self.wt7
    * self.hn1*(1-self.hn1) * self.x2
```

```
self.dbs1 = (self.hn4-y) * self.hn4*(1-self.hn4) * self.wt7
* self.hn1*(1-self.hn1)

self.dwt3 = (self.hn4-y) * self.hn4*(1-self.hn4) * self.wt8
* self.hn2*(1-self.hn2) * self.x1
self.dwt4 = (self.hn4-y) * self.hn4*(1-self.hn4) * self.wt8
* self.hn2*(1-self.hn2) * self.x2
self.dbs2 = (self.hn4-y) * self.hn4*(1-self.hn4) * self.wt8
* self.hn2*(1-self.hn2)

self.dwt5 = (self.hn4-y) * self.hn4*(1-self.hn4) * self.wt9
* self.hn3*(1-self.hn3) * self.x1
self.dwt6 = (self.hn4-y) * self.hn4*(1-self.hn4) * self.wt9
* self.hn3*(1-self.hn3) * self.x2
self.dbs3 = (self.hn4-y) * self.hn4*(1-self.hn4) * self.wt9
* self.hn3*(1-self.hn3)

def fit(self, X, Y, epochs=1, learning_rate=1,
initialise=True, display_loss=False):

  # initialise w, b
  if initialise:
    self.wt1 = np.random.randn()
    self.wt2 = np.random.randn()
    self.wt3 = np.random.randn()
    self.wt4 = np.random.randn()
    self.wt5 = np.random.randn()
    self.wt6 = np.random.randn()
    self.wt7 = np.random.randn()
    self.wt8 = np.random.randn()
    self.wt9 = np.random.randn()
    self.bs1 = 0
    self.bs2 = 0
```

```
    self.bs3 = 0
    self.bs4 = 0

if display_loss:
  loss = {}

for i in tqdm_notebook(range(epochs), total=epochs,
unit="epoch"):
  dwt1, dwt2, dwt3, dwt4, dwt5, dwt6, dwt7, dwt8, dwt9,
  dbs1, dbs2, dbs3, dbs4 = [0]*13
  for x, y in zip(X, Y):
    self.grad(x, y)
    dwt1 += self.dwt1
    dwt2 += self.dwt2
    dwt3 += self.dwt3
    dwt4 += self.dwt4
    dwt5 += self.dwt5
    dwt6 += self.dwt6
    dwt7 += self.dwt7
    dwt8 += self.dwt8
    dwt9 += self.dwt9
    dbs1 += self.dbs1
    dbs2 += self.dbs2
    dbs3 += self.dbs3
    dbs4 += self.dbs4

  m = X.shape[1]
  self.wt1 -= learning_rate * dwt1 / m
  self.wt2 -= learning_rate * dwt2 / m
  self.wt3 -= learning_rate * dwt3 / m
  self.wt4 -= learning_rate * dwt4 / m
  self.wt5 -= learning_rate * dwt5 / m
  self.wt6 -= learning_rate * dwt6 / m
```

```
    self.wt7 -= learning_rate * dwt7 / m
    self.wt8 -= learning_rate * dwt8 / m
    self.wt9 -= learning_rate * dwt9 / m
    self.bs1 -= learning_rate * dbs1 / m
    self.bs2 -= learning_rate * dbs2 / m
    self.bs3 -= learning_rate * dbs3 / m
    self.bs4 -= learning_rate * dbs4 / m

    if display_loss:
      Y_pred = self.predict(X)
      loss[i] = mean_squared_error(Y_pred, Y)

  if display_loss:
    plt.plot(loss.values())
    plt.xlabel('Epochs')
    plt.ylabel('Mean Squared Error')
    plt.show()

def predict(self, X):
  #predicting the results on unseen data
  Y_pred = []
  for x in X:
    y_pred = self.forward_pass(x)
    Y_pred.append(y_pred)
  return np.array(Y_pred)

ffn = simpleFFNetwork()

#train the model on the data
ffn.fit(X_train, Y_train, epochs=2000, learning_rate=.01,
display_loss=True)

#predictions
Y_pred_train = ffn.predict(X_train)
```

```
Y_pred_binarised_train = (Y_pred_train >= 0.5).
astype("int").ravel()
Y_pred_val = ffn.predict(X_val)
Y_pred_binarised_val = (Y_pred_val >= 0.5).
astype("int").ravel()
accuracy_train = accuracy_score(Y_pred_binarised_train,
Y_train)
accuracy_val = accuracy_score(Y_pred_binarised_val, Y_val)

#model performance
print("Training accuracy", round(accuracy_train, 2))
print("Validation accuracy", round(accuracy_val, 2))

#visualize the predictions
plt.scatter(X_train[:,0], X_train[:,1], c=Y_pred_binarised_
train, cmap=my_cmap, s=15*(np.abs(Y_pred_binarised_train-Y_
train)+.2))
plt.show()
```

Here's the output:

```
(500, 2) (500,)
(375, 2) (125, 2)
```

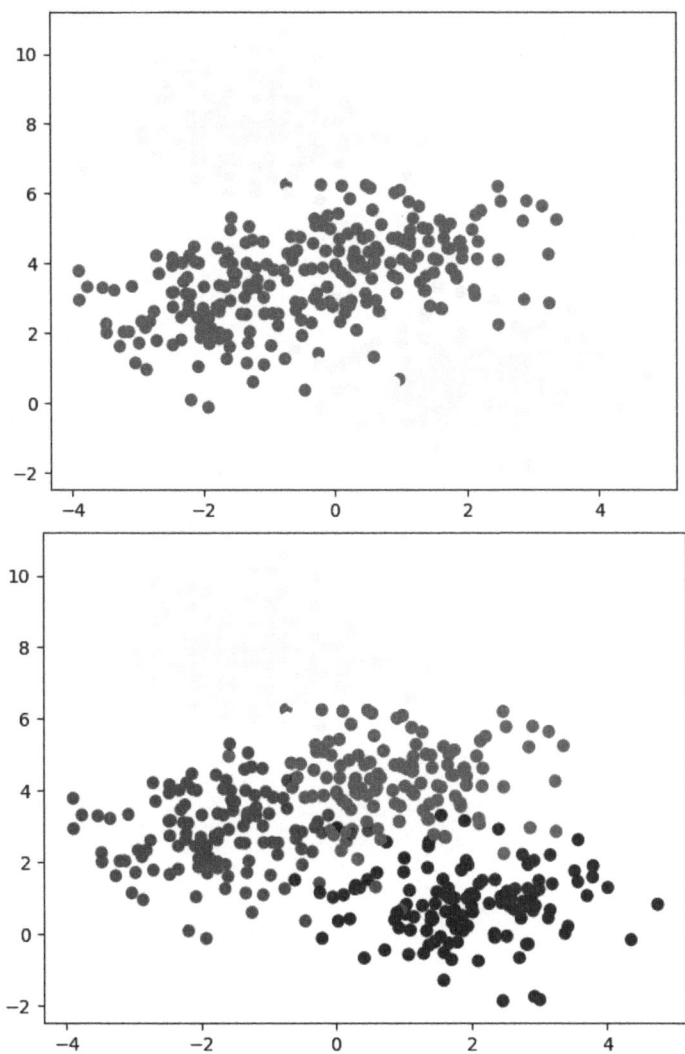

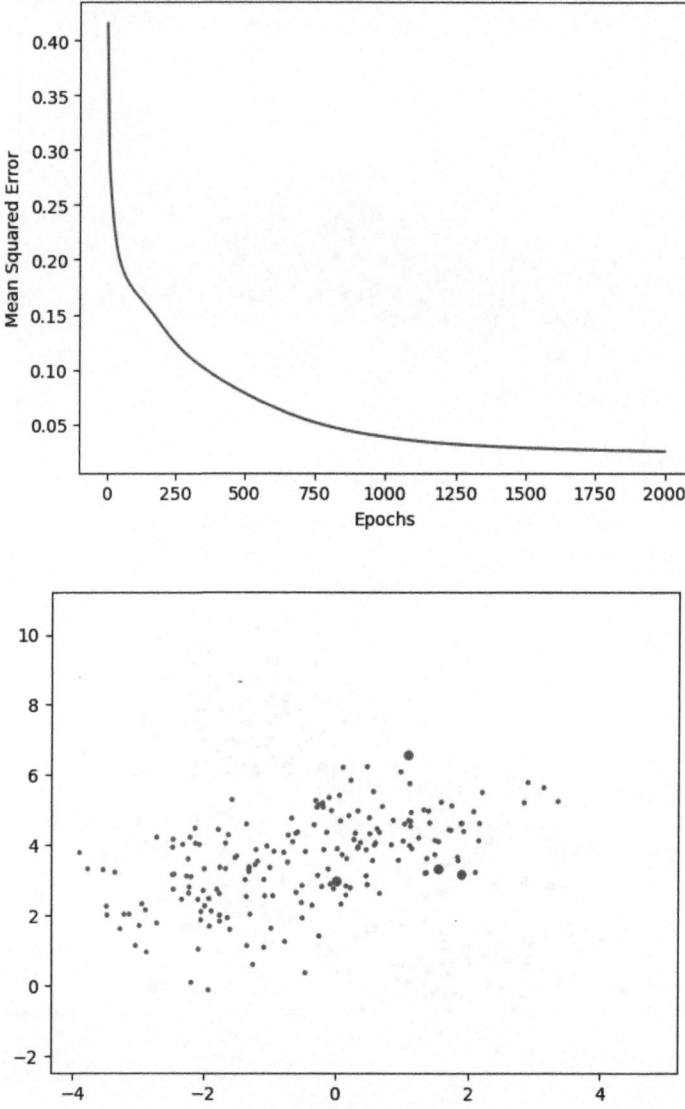

Figure 6-4. *Illustration of multilayer feed-forward network*

6.5.2 Feedback Networks

The last three types of ANN (single node with its own feedback, single-layer recurrent, and multilayer recurrent) are categorized under feedback ANN in which the interconnections allow feedback in the reverse direction. There exists a feedback interconnection from the output layer to the hidden or output layer.

Single node with its own feedback
The interconnection of the neural network with a feedback connection from its output back to its input is called *single node* with its feedback network. This has only a single node with its feedback interconnection as the network architecture, as illustrated in Figure 6-5.

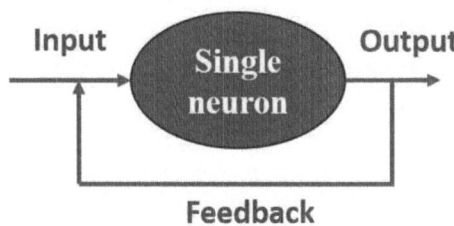

Figure 6-5. *Single-node feedback network*

Single-layer recurrent network
The interconnection of this network is similar to a single-node feedback network. In this type, one single layer (outputs) is receiving its own feedback and feedback from other neurons of the output layer, as depicted in Figure 6-6. The initial weights are assigned to the interconnections from the inputs to the output neurons.

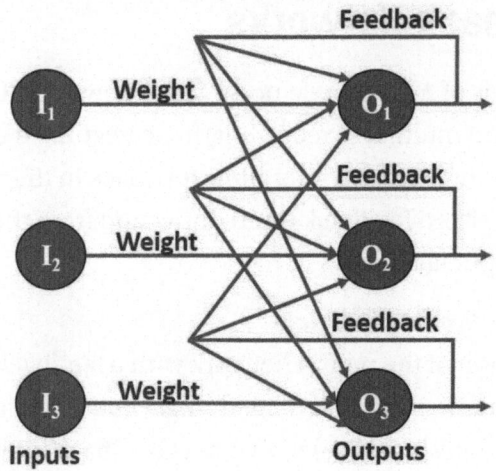

Figure 6-6. *Single-layer recurrent network*

Multilayer recurrent network

```
import numpy as np
import matplotlib.pyplot as plt

# Generate sine wave data
data = np.cos(np.linspace(0, 100, 100))  # 100 data points

# Prepare the input and output sequences
n_steps = 10
X = np.array([data[i:i+n_steps] for i in range(len(data)-
n_steps)])
y = np.array([data[i+n_steps] for i in range(len(data)-
n_steps)])

# RNN parameters
input_size = 1
hidden_size = 10
output_size = 1
learning_rate = 0.01
```

```
# Initialize weights and biases
Wx = np.random.randn(hidden_size, input_size)
# Input to hidden
Wh = np.random.randn(hidden_size, hidden_size)
# Hidden to hidden
Wy = np.random.randn(output_size, hidden_size)
# Hidden to output
bh = np.zeros((hidden_size, 1))   # Hidden bias
by = np.zeros((output_size, 1))   # Output bias

# signmoid activation function
def sigmoid(x):
    return (1 / (1 + np.exp(-x)))

# Forward pass
def forward(inputs):
    xh = np.zeros((hidden_size, 1))
    outputs = []
    for x in inputs:
        x = x.reshape(-1, 1)  # Convert to column vector
        mulx = np.dot(Wx, x)
        mulh = np.dot(Wh, xh)
        xh = mulx + mulh + bh
        xh = sigmoid(xh)
         #weight vector is multiplied with the vector h to give
         the output vector
        y = Wy @ xh + by
        outputs.append(y)
    return outputs, xh

# Training loop
for epoch in range(100):  # 100 epochs
    for i in range(len(X)):
```

```
        inputs = X[i].reshape(n_steps, input_size)
        target = y[i].reshape(output_size, 1)

        # Forward pass
        outputs, h = forward(inputs)

        # Compute error
        error = outputs[-1] - target

        # Backpropagation through time (simplified)
        dWy = error @ h.T
        dWh = ((1 - h**2) * (Wy.T @ error)) @ h.T
        dWx = ((1 - h**2) * (Wy.T @ error)) @ inputs[-1].
        reshape(1, -1).T

        # Update weights and biases
        Wy -= learning_rate * dWy
        Wh -= learning_rate * dWh
        Wx -= learning_rate * dWx
        by -= learning_rate * error
        bh -= learning_rate * (1 - h**2) * (Wy.T @ error)

# Prediction
predictions = []
for i in range(len(X)):
    inputs = X[i].reshape(n_steps, input_size)
    outputs, _ = forward(inputs)
    predictions.append(outputs[-1].item())

# Plot the results
plt.plot(data, label='Original Data')
plt.plot(range(n_steps, len(predictions) + n_steps),
predictions, label='Predictions', color='r')
plt.legend()
plt.show()
```

Here is the output:

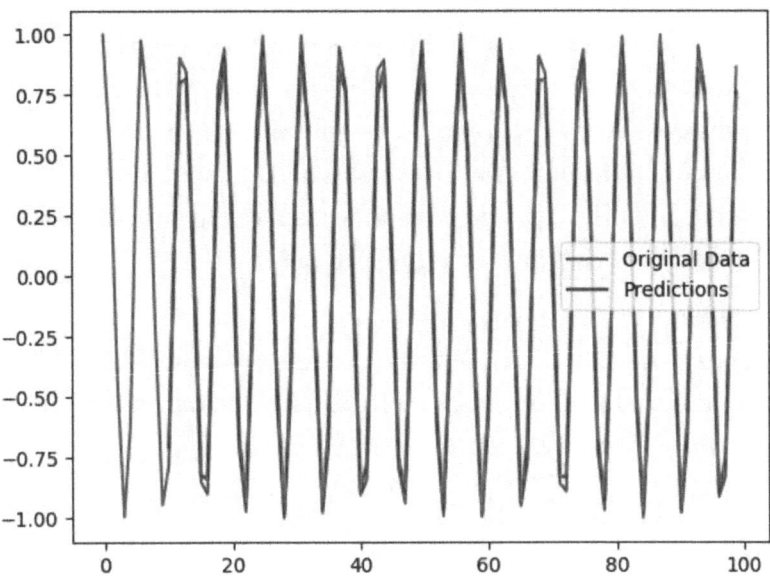

6.5.3 Unsupervised ANNs

Unsupervised ANNs explore the relationship between inputs and categorize them into unlabeled groups. Unsupervised algorithms in ANNs include restricted Boltzmann machines, self-organized maps, and autoencoders.

Restricted Boltzmann machines (RBM)

The two-layer undirected network comprises only two layers: visible layer and hidden layer. The visible layer represents the input data, and the hidden layer studies the features of the input.

It is mainly used in feature learning, topic modeling, etc. It is also used as a building block in deep belief networks.

Self-organizing maps (SOMs)

It is an unsupervised learning to generate the 2D representation of input data called *maps* like dimensionality reduction. SOM works in a different way unlike error correction learning. It uses competitive learning, which involves the competing neurons and neighborhood function to decide the winner neuron. Consequently, the weights of the neighborhood neurons are updated, and the map is grown and reshaped. The process is repeated until the inputs are categorized in the best way.

Autoencoders

Autoencoders consist of encoders and decoders that map the input space to a lower-dimensional intermediate representation and regenerate input data from the intermediate representation, respectively.

6.6 Summary

Artificial neural networks are imitations of the human brain's neural networks, which are used in predictions. The components involved are input, hidden, and output layers; weights; bias; activation functions; and loss functions. Two major types of ANNs are feed-forward and feedback ANNs. We discussed both types and their subtypes in this chapter. Also, the Python programs for implementing activation functions, loss functions, feed-forward, and feedback network architectures were discussed. In the next chapter, we will discuss the convolutional neural networks in more detail.

CHAPTER 7

Convolutional Neural Networks for IoT

7.1 Introduction

Deep neural networks have evolved into various types that are suitable for different domains. Convolutional neural networks (CNNs) are one type that are the right fit for computer vision applications. Object detection and image processing applications can be developed using CNNs. CNNs are different from other machine learning techniques because they can extract features on their own from images as well as other types of data without manual operations involved. There are many built-in CNNs such as AlexNet, GoogleNet, ResNet, etc.

A sound implementation of a CNN lies in the better understanding and usage of its basic operations and the components that it consists of. The basic components are kernels, convolution operation, feature map, pooling, and striding.

- **Kernel/filter:** This is a small matrix-like grid that moves over the image, which is represented as a bigger matrix. This results in identifying almost all the required patterns from the image such as lines, shapes, etc. The size of the kernel is usually smaller than the input image size.

- **Convolution operation:** The kernel size is a hyper parameter in the convolution operation, which is generally an odd-numbered matrix such as 3x3, 5x5, or 7x7. The kernel slides over the input image and calculates the sum of multiplications performed element-wise between the kernel and the portion of the input image covered by the kernel during the movement. Such calculated single values during the movement are framed by positioning the resulting sum on the cell of the new feature map matrix similar to the kernel placed over the input image.

- **Feature map:** This is the output from the convolution layer that carries the essential features from the input or previous layer. The filter matrices are multiplied with the image to produce feature maps. Each filter results in a different feature map for extracting specific features of the image. Figure 7-1 shows the generation of feature maps using three parameters: stride, depth, and padding. Stride is the number of pixels skipped during the movement of filtering over the image after a multiplication operation between the elements of the filter and the pixels it covers. Depth refers to the number of filters that are used to generate the same number of feature maps. Padding adds pixels along the borders of the image to calculate the value of the border pixels in generating feature maps.

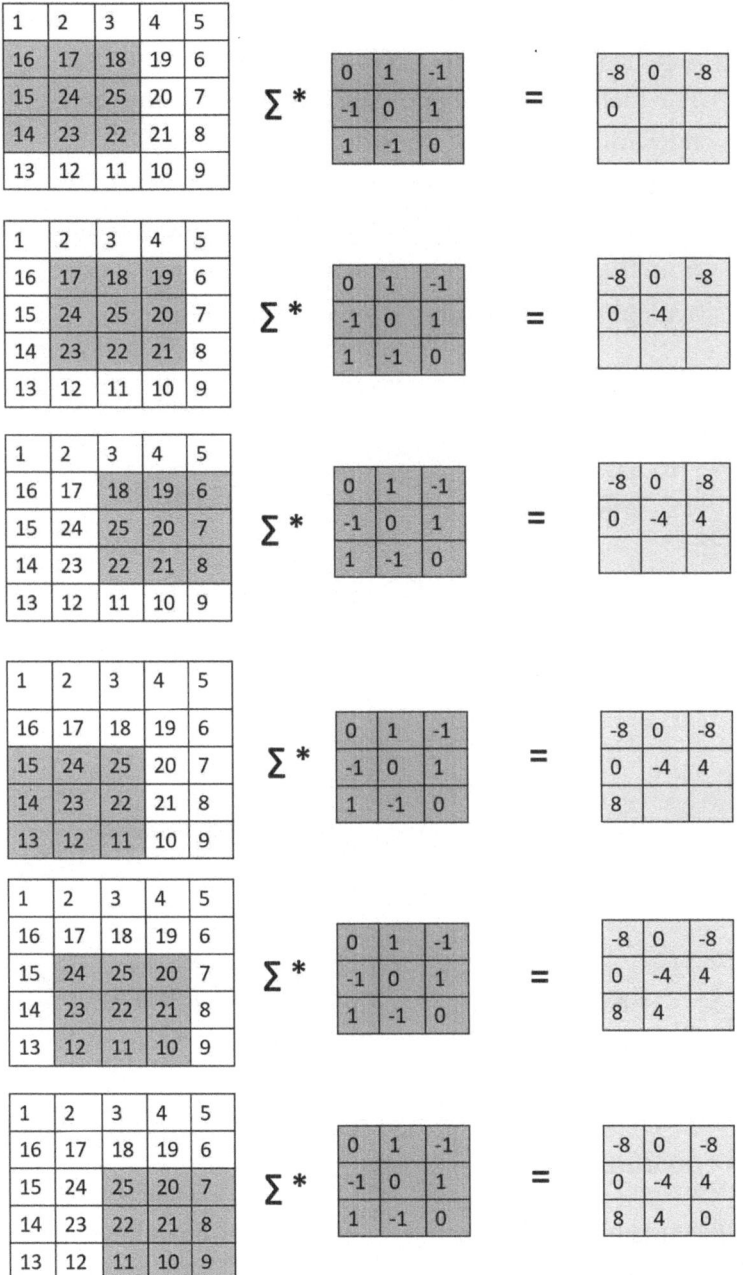

Figure 7-1. *Generation of 3x3 feature maps with 5x5 input image size, 3x3 filter size, and stride = 1*

- **Pooling:** This means downsampling of feature maps by aggregating the subset of elements in the feature map. The main objective of pooling is to preserve the most relevant information while downsizing the size of the feature map. It comes in two types, as demonstrated in Figure 7-2. Max pooling takes the maximum value from the pooling region, and average pooling takes the average value of all pixels in the pooling region.

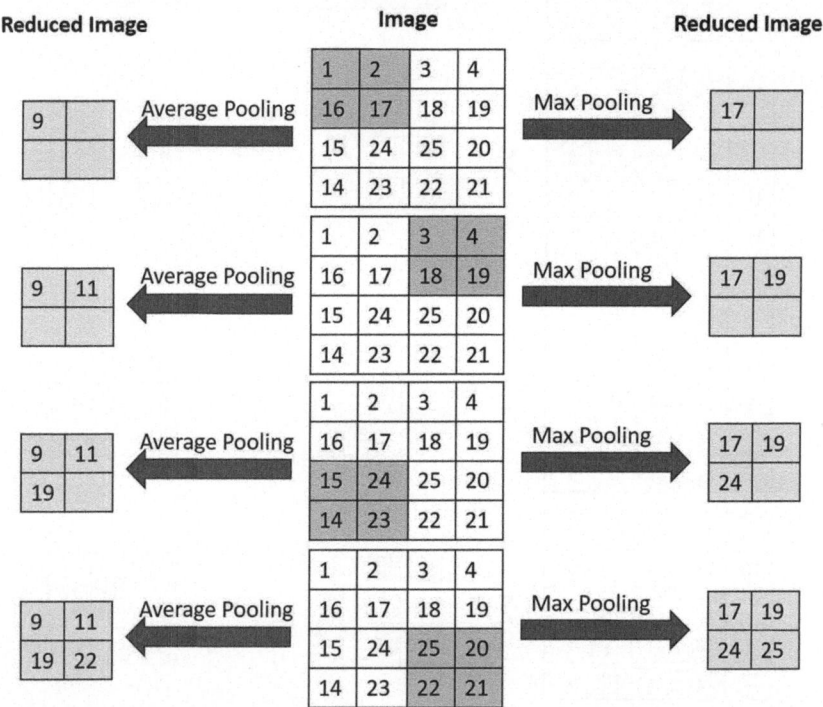

Figure 7-2. *Demonstration of two major pooling: average pooling and max pooling*

The formula for calculating the output dimension of the convolution layer is as follows:

$$W_o = ((W_i - F + 2P)/S) + 1 \qquad \text{-------- Eqn 1}$$

where *Wi* and *Wo* are the input image and the output feature map size in pixel values. *F* is the size of the filter or kernel. *P* is the amount of padding for extending the borders of input image. *S* is the number of rows or columns skipped during the movement of the filter over the input mage. Hence, after applying convolution, *Wi x Wi x Din* get transformed into *Wo x Wo x Dout.*

7.2 General Architecture of CNN

The general architecture of the CNN has three main layers: the convolution layer, pooling layer, and classification layer. But before all these main layers, the input layer has the image in a prescribed size for all the successive operations.

Input Layer

The input layer is the layer that has the input image with a predefined size. The standard dimension of the image is 227x227x3 or 224x224x3 as an RGB image. This image dimension is followed in almost all types of CNN. Alternatively, the grayscale image dimension is 32x32x1.

Convolution Layer

The process of CNN starts with the convolution layer in which the convolution operation is carried out between the input image and the filter or kernel. The common filter sizes are 2x2, 3x3, 5x5, and 7x7. The filter slides over the input image, and as it slides, the dot product is calculated between the portion of the image covered by filter and the filter itself.

This results in a feature map that provides details about the image such as edges, corners, notable parts of the objects, etc. The convolution layer can generate a greater number of feature maps to provide various details of the image with the usage of multiple filters. The feature maps in turn feed to successive layers to learn more and more about the features of the input image. The positivity of the convolution layer is that it preserves the spatial relationships between pixels.

Pooling Layer

This layer acts as a bridge between the convolution layer and the fully connected layer. The key role of this layer is the reduce the dimension of the feature maps to reduce the computation cost. This pooling process is performed independently on each feature map irrespective of any order in the feature maps. This layer along with the convolution layer works for extracting the features and helps in recognizing the features independently.

Fully Connected Layer

This layer serves the completely connected neural network layer with weights along the connections and biases between layers. These layers form the last layers of the overall CNN architecture. The output from the previous layer is processed and flattened to form the fully connected (FC) layer. Here the regular mathematical functions of neural networks happens, and the classification process commences. Usually, more than one layer is connected in the FC layer to get better results. Also, this layer involves no human intervention, monitoring, or control.

Apart from the three layers, dropout and activation functions are added to improve the learning process and accuracy of the result. There are situations in which the fully connected layer leads to overfitting during training. This may result in under performance when the model is used for new data. To handle overfitting, a dropout layer is used to drop some neurons randomly. When the dropout layer is set to 0.2, then 20% of the neurons are dropped from the fully connected layer.

Another parameter that contributes to the CNN model is the activation function. In fact, all neural network models use the activation functions. Activation functions make the decision about which neurons need to be fired and forwarded to the end of the network. Various activation functions are available for different types of classification. Activation functions are required to improve the model by providing nonlinearity over the network. Sigmoid, tanH, ReLU, and softmax are the commonly used activation functions in neural networks.

7.3 Types of CNNs

Many types of CNNs are defined, and each type is best when it is published with results to the problems it addressed. They are LeNet, AlexNet, VGGNet, GoogLeNet, and ResNet.

LeNet

This type of CNN is mainly focused on handwritten character recognition with the regular convolution, pooling, and fully connected layers that have nonlinear activation functions. The convolution layer is for extracting features, pooling for downsampling, and using tanh activation functions for nonlinearity. The FC layer is for classifying the input images under various classes. To avoid overfitting of the model, dropout is used to reduce complexity and improve performance.

The architecture is demonstrated in Figure 7-3 as it consists of three convolution layers and two pooling layers alternatively. Also, it comprises three fully connected layers with a flattened size of 120, 84, and 10, respectively. The output is categorized under any one of the 10 classes.

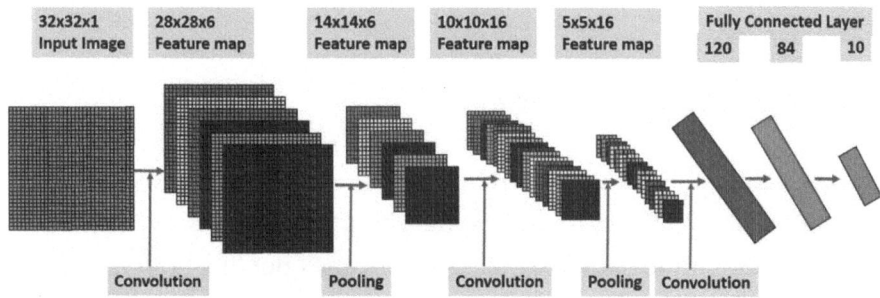

Figure 7-3. *LeNet architecture*

AlexNet

This model was proposed in 2012 through a research paper by Alex Krizhevsky and his fellow researchers. It is a deep architecture that involves five convolution layer along with max pooling. As we walk through the architecture of AlexNet as shown in Figure 7-4, the input image of size

215

227x227x3 is fed to the first layer of convolution with a 11x11 filter size with stride 4. So, 96 filters are applied in the first layer over the input layer. The output dimension of the first layer is 55x55x96. The second layer is the max pooling layer, which reduces the dimension as 27x27x96.

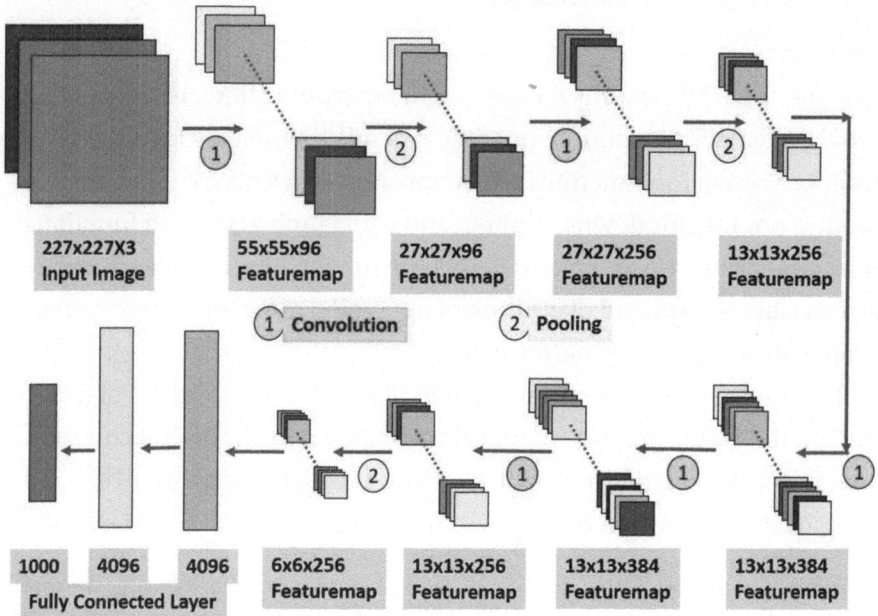

Figure 7-4. *AlexNet architecture*

The second convolution operation is performed with a filter size of 5x5. The number of filters are 256 along with a padding of size 2 and stride 1. The resultant dimension is 27x27x256. The max pooling layer is again applied and downsized. The feature map has the dimension 13x13x256 with stride 2 and pooling with a size of 3x3.

The third convolution operation is performed with 384 filters of size 3x3 with stride 1 and padding 1. The output dimension is 13x13x384. The fourth convolution operation with the same dimension is performed, and the result is 13x13x384. The final convolution layer is with 3x3 filters with 256 numbers. The resulting dimension is 13x13x256. Next, the third max pooling layer is applied, and the dimension is reduced to 6x6x256.

216

Later, the first fully connected layer of output size 4096 is generated. Then one more fully connected layer is formed. Finally, the output layer is formed with a fully connected 1,000 classes.

As we move forward through the model, the number of filters increases, and it helps to extract features in more detail in the architecture. Also, the feature map size decreases as we move deeper.

VGGNet

The VGGNet model comprises 13 convolutional layers and 3 fully connected layers along with pooling layers. The input layer is fed to the first convolution layer, which has 64 filters and of size 3x3. This is repeated one more time in the second convolution layer. The output of these two layers is 224x224x64. Now, max pooling is applied with a 2x2 size with stride 2, and the output dimension is 112x112x128.

Next, two more convolution layers are formed with 128 filters of size 3x3. The new dimension in these layers is 112x112x128. Max pooling is applied, and the resulting dimension is 56x56x256.

Next, three more convolution layers are formed with 256 filters of size 3x3. The dimension in these layers is reduced to 28x28x512. Max pooling is applied, and the resulting dimension is 14x14x512. The same set of three convolution and one max pooling is repeated to produce a 7x7x512 feature map. Finally, it is flattened into three fully connected layers with two 4,096 sizes and 1,000 categorized classes.

The architecture shown in Figure 7-5 is VGGNet 16 as explained. Similar to this is VGGNet 19, which has three more convolution layers.

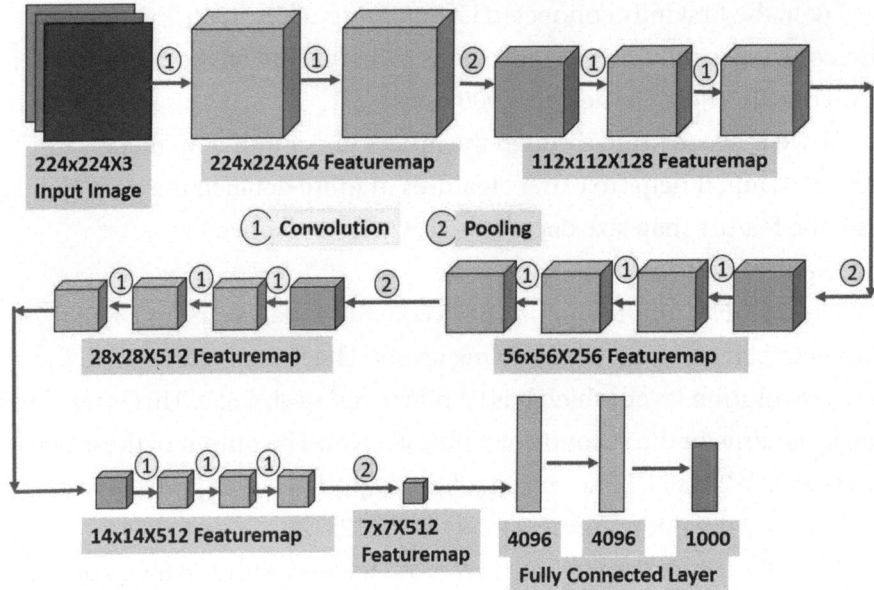

Figure 7-5. *VGGNet 16 architecture*

GoogLeNet

This CNN model is very different from other models as it uses 1x1 convolution, global average pooling, inception module, and auxiliary classifier for training.

Global average pooling: This layer transforms the 7x7 feature map to 1x1, which highly decreases the computation cost. Other networks such as AlexNet and ZFNet have fully connected layer loaded with more parameters, which obviously increases the computation cost.

1x1 convolution: This convolution operation is used in the inception module. It is used to reduce the number of weights and biases of the architecture to increase the depth of the architecture.

Inception module: In other models, the convolution layer is of fixed size, but in this inception module three different size convolutions along with max pooling is performed in order to generate final output concatenated from the four previously mentioned operations. As the

218

convolution is performed with different sizes parallelly along with max pooling, it will provide better understanding of the features and hence contribute to improved classification.

Auxiliary classifier for training: This part of network is used for training only with 5x5 average pooling, 1x1 convolution with 128 filters, 2 fully connected layer of size 1024, and finally 1,000 outputs along with the softmax layer. These are said to be intermediate classifier branches that are used to oppose against the gradient vanishing problem and to offer regularization.

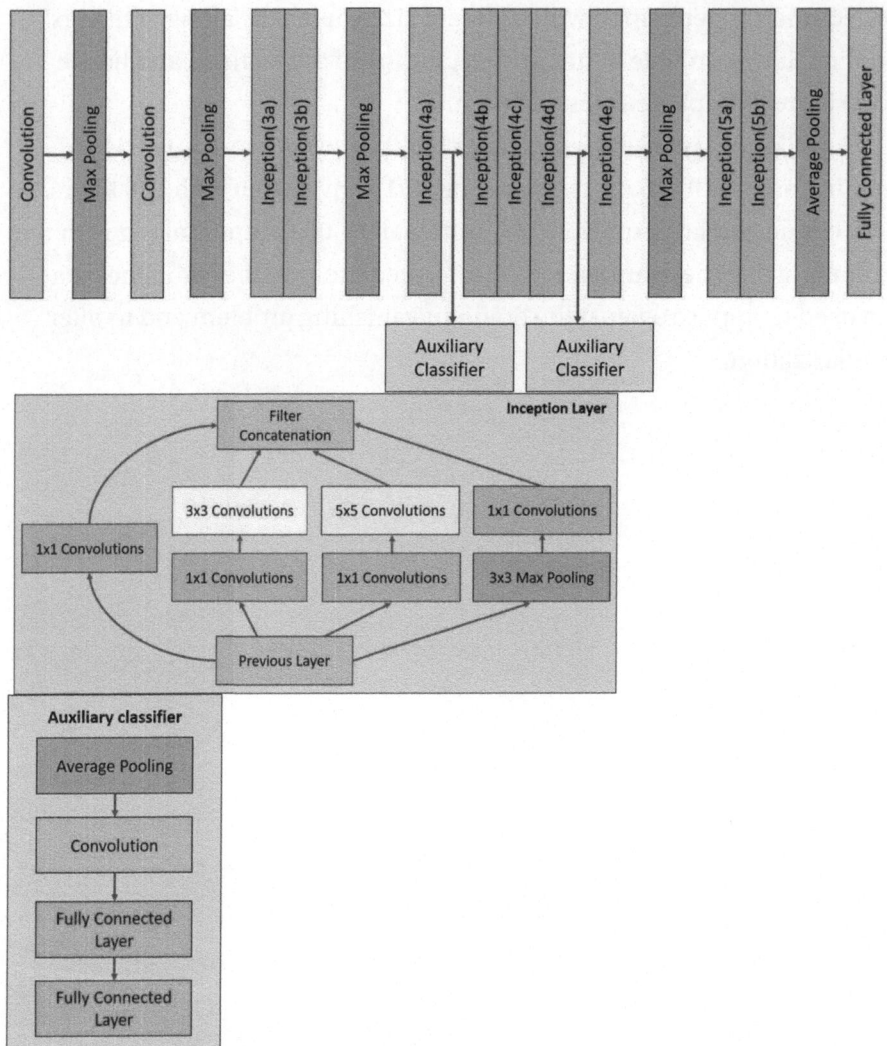

Figure 7-6. *GoogLeNet architecture*

ResNet

The ResNet model starts with an input image size of 227x227x3. The next is zero padding, which increases the dimension by 229x229x3. This is followed by the convolution layer with a kernel size of 7x7. The resultant dimension is 112x112x64 feature maps. Next, max pooling is applied to reduce the size. Now the dimension is 56x56x64.

220

Now, the actual ResNet layer starts with a convolution 2x operation, which consists of six convolution layers with the dimension 56x56x64. Next the convolution 3x is formed with eight convolution layers of dimension 28x28x128 feature maps.

Later, the convolution 4x is formed with 12 convolution layers of dimension 14x14x 256. Finally, the convolution 5x is formed with six convolution layers of dimension 7x7x512. Later the max pool is applied and flattened to a single fully connected layer of size 1000. Figure 7-7 shows the demonstration of ResNet 34. Similarly, ResNet 50, ResNet 101, etc., are built with the basic convolution, pooling, and fully connected layers.

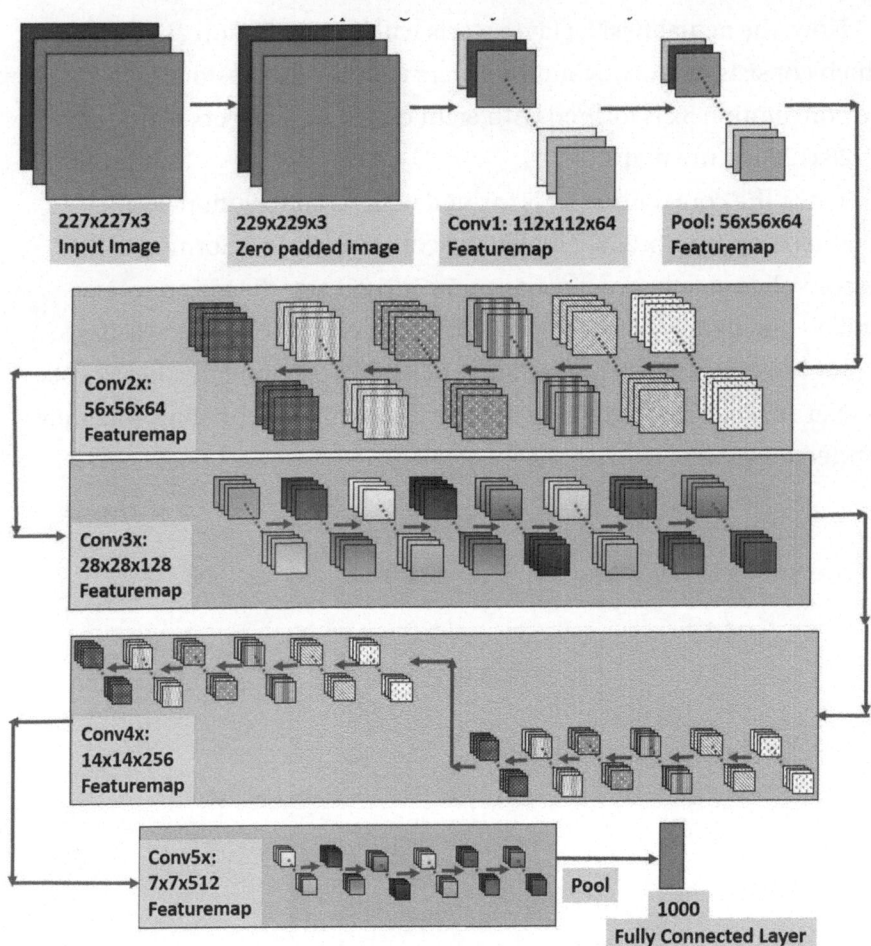

Figure 7-7. *ResNet 34 architecture*

7.4 Case Study for Computer Vision

"""CNN - Fashion MNIST example - Full code.ipynb

"""

```
# Importing the necessary packages
import tensorflow as tf
```

```
import tensorflow_datasets as tfds
import datetime
from google.colab import drive
drive.mount('/content/drive')
import os
os.chdir("drive/My Drive/Colab Notebooks/LAIoT/Chapter 7")
"""## Downloading and preprocessing the data"""
# Before continuing with building the model and training, the
main part is to preprocess the dataset.
# This is a very important step in all of machine learning
# The Fashion MNIST dataset is, in general, highly processed
already - after all its 28x28 grayscale images of clearly
visible digits
# Thus, our preprocessing will be limited to scaling the pixel
values, shuffling the data and creating a validation set
# Defining some constants/hyperparameters
BUFFER_SIZE = 70_000 # for reshuffling
BATCH_SIZE = 128
NUM_EPOCHS = 10
# Downloading the MNIST dataset
# When 'with_info' is set to True, tfds.load() returns two
variables:
# - the dataset (including the train and test sets)
# - meta info regarding the dataset itself
mnist_dataset, mnist_info = tfds.load(name='fashion_mnist',
with_info=True, as_supervised=True)
# Extracting the train and test datasets
mnist_train, mnist_test = mnist_dataset['train'],
mnist_dataset['test']
# Creating a function to scale our image data (it is
recommended to scale the pixel values in the range [0,1] )
```

```python
def scale(image, label):
    image = tf.cast(image, tf.float32)
    image /= 255.
    return image, label
# Scaling the data
train_and_validation_data = mnist_train.map(scale)
test_data = mnist_test.map(scale)
# Defining the size of the validation set
num_validation_samples = 0.1 * mnist_info.splits['train'].
num_examples
num_validation_samples = tf.cast(num_validation_samples,
tf.int64)
# Defining the size of the test set
num_test_samples = mnist_info.splits['test'].num_examples
num_test_samples = tf.cast(num_test_samples, tf.int64)
# Reshuffling the dataset
train_and_validation_data = train_and_validation_data.
shuffle(BUFFER_SIZE)
# Splitting the dataset into training + validation
train_data = train_and_validation_data.skip(num_validation_
samples)
validation_data = train_and_validation_data.take(num_
validation_samples)
# Batching the data
# NOTE: For proper functioning of the model, we need to create
one big batch for the validation and test sets
train_data = train_data.batch(BATCH_SIZE)
validation_data = validation_data.batch(num_validation_samples)
test_data = test_data.batch(num_test_samples)
"""## Creating the model and training it"""
# Now that we have preprocessed the dataset, we can define our
```

```
CNN and train it
# Outlining the model/architecture of our CNN
# CONV -> MAXPOOL -> CONV -> MAXPOOL -> FLATTEN -> DENSE
model = tf.keras.Sequential([
    tf.keras.layers.Conv2D(50, 5, activation='relu',
input_shape=(28, 28, 1)),
    tf.keras.layers.MaxPooling2D(pool_size=(2,2)),
    # (2,2) is the default pool size so we could have just used
MaxPooling2D() with no explicit arguments
    tf.keras.layers.Conv2D(50, 3, activation='relu'),
    tf.keras.layers.MaxPooling2D(pool_size=(2,2)),
    tf.keras.layers.Flatten(),
    tf.keras.layers.Dense(10) # You can apply softmax
activation here, see below for comentary
])
# A brief summary of the model and parameters
model.summary(line_length = 75)

#ouput


Model: "sequential"
```

Layer (type)	Output Shape	Param #
conv2d (Conv2D)	(None, 24, 24, 50)	1300
max_pooling2d (MaxPooling2D)	(None, 12, 12, 50)	0
conv2d_1 (Conv2D)	(None, 10, 10, 50)	22550
max_pooling2d_1 (MaxPooling2D)	(None, 5, 5, 50)	0
flatten (Flatten)	(None, 1250)	0
dense (Dense)	(None, 10)	12510

Total params: 36360 (142.03 KB)
Trainable params: 36360 (142.03 KB)
Non-trainable params: 0 (0.00 Byte)

```
# Defining the loss function
# In general, our model needs to output probabilities of
each class,
# which can be achieved with a softmax activation in the last
dense layer
# However, when using the softmax activation, the loss can
rarely be unstable
# Thus, instead of incorporating the softmax into the
model itself,
# we use a loss calculation that automatically corrects for the
missing softmax
# That is the reason for 'from_logits=True'
loss_fn = tf.keras.losses.SparseCategoricalCrossentropy(fr
om_logits=True)
# Compiling the model with Adam optimizer and the cathegorical
crossentropy as a loss function
model.compile(optimizer='adam', loss=loss_fn,
metrics=['accuracy'])
# Defining early stopping to prevent overfitting
early_stopping = tf.keras.callbacks.EarlyStopping(
    monitor = 'val_loss',
    mode = 'auto',
    min_delta = 0,
    patience = 2,
    verbose = 0,
    restore_best_weights = True
)
```

```
# Logging the training process data to use later in tensorboard
log_dir = "logs\\fit\\" + datetime.datetime.now().
strftime("%Y%m%d-%H%M%S")
tensorboard_callback = tf.keras.callbacks.TensorBoard(log_
dir=log_dir, histogram_freq=1)
# Train the network
model.fit(
    train_data,
    epochs = NUM_EPOCHS,
    callbacks = [tensorboard_callback,early_stopping],
    validation_data = validation_data,
    verbose = 2
)
```

#output

Epoch 1/10
422/422 - 66s - loss: 0.6083 - accuracy: 0.7824 - val_loss:
0.4082 - val_accuracy: 0.8518 - 66s/epoch - 157ms/step
Epoch 2/10
422/422 - 60s - loss: 0.3943 - accuracy: 0.8609 - val_loss:
0.3625 - val_accuracy: 0.8748 - 60s/epoch - 141ms/step
Epoch 3/10
422/422 - 61s - loss: 0.3484 - accuracy: 0.8777 - val_loss:
0.3344 - val_accuracy: 0.8772 - 61s/epoch - 145ms/step
Epoch 4/10
422/422 - 61s - loss: 0.3184 - accuracy: 0.8876 - val_loss:
0.2932 - val_accuracy: 0.8978 - 61s/epoch - 144ms/step
Epoch 5/10
422/422 - 59s - loss: 0.3003 - accuracy: 0.8925 - val_loss:
0.2778 - val_accuracy: 0.9072 - 59s/epoch - 139ms/step
Epoch 6/10

```
422/422 - 60s - loss: 0.2806 - accuracy: 0.9004 - val_loss:
0.2690 - val_accuracy: 0.9015 - 60s/epoch - 142ms/step
Epoch 7/10
422/422 - 59s - loss: 0.2671 - accuracy: 0.9032 - val_loss:
0.2480 - val_accuracy: 0.9108 - 59s/epoch - 141ms/step
Epoch 8/10
422/422 - 61s - loss: 0.2539 - accuracy: 0.9083 - val_loss:
0.2483 - val_accuracy: 0.9117 - 61s/epoch - 144ms/step
Epoch 9/10
422/422 - 59s - loss: 0.2451 - accuracy: 0.9119 - val_loss:
0.2295 - val_accuracy: 0.9128 - 59s/epoch - 140ms/step
Epoch 10/10
422/422 - 58s - loss: 0.2364 - accuracy: 0.9159 - val_loss:
0.2379 - val_accuracy: 0.9188 - 58s/epoch - 139ms/step
<keras.src.callbacks.History at 0x7f759dd6f070>
"""## Testing our model"""
# Testing our model
test_loss, test_accuracy = model.evaluate(test_data)

#output
1/1 [==========================] - 5s 5s/step - loss: 0.2767 -
accuracy: 0.9014
# Printing the test results
print('Test loss: {0:.4f}. Test accuracy: {1:.2f}%'.
format(test_loss, test_accuracy*100.))

#output
Test loss: 0.2767. Test accuracy: 90.14%
"""### Plotting images and the results"""
import matplotlib.pyplot as plt
import numpy as np
# Split the test_data into 2 arrays, containing the images and
the corresponding labels
```

```
for images, labels in test_data.take(1):
    images_test = images.numpy()
    labels_test = labels.numpy()
# Reshape the images into 28x28 form, suitable for matplotlib
(original dimensions: 28x28x1)
images_plot = np.reshape(images_test, (10000,28,28))
# prompt: create a dataframe with 2 columns and 10 rows
import pandas as pd
data = {'column1': [0, 1, 2, 3, 4, 5, 6, 7, 8, 9],
       'column2': ['T-shirt/top', 'Trouser', 'Pullover',
       'Dress', 'Coat','Sandal', 'Shirt', 'Sneaker', 'Bag',
       'Ankle boot']}
df = pd.DataFrame(data)
print(df)
```

#output

	column1	column2
0	*0*	*T-shirt/top*
1	*1*	*Trouser*
2	*2*	*Pullover*
3	*3*	*Dress*
4	*4*	*Coat*
5	*5*	*Sandal*
6	*6*	*Shirt*
7	*7*	*Sneaker*
8	*8*	*Bag*
9	*9*	*Ankle boot*

```
# The image to be displayed and tested
i = 600
# Plot the image
plt.figure(figsize=(3, 3))
```

```
plt.axis('off')
plt.imshow(images_plot[i-1], cmap="gray", aspect='auto')
plt.show()
# Print the correct label for the image
print("Label: {} {}".format(labels_test[i-1], df.
column2[labels_test[i-1]]))
```

#output

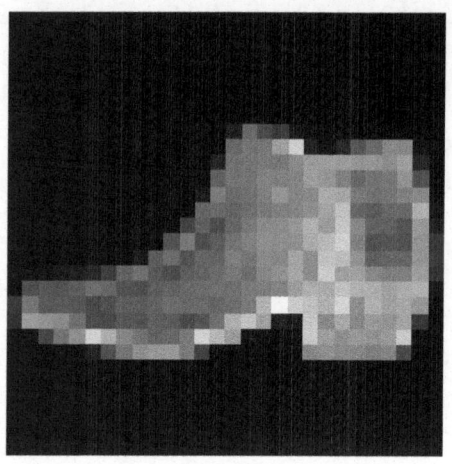

Label 9: Ankle boot

```
# Obtain the model's predictions (logits)
predictions = model.predict(images_test[i-1:i])
# Convert those predictions into probabilities (recall that we
incorporated the softmaxt activation into the loss function)
probabilities = tf.nn.softmax(predictions).numpy()
# Convert the probabilities into percentages
probabilities = probabilities*100
# Create a bar chart to plot the probabilities for each class
plt.figure(figsize=(12,5))
```

```
plt.bar(x=[0,1,2,3,4,5,6,7,8,9], height=probabilities[0],
tick_label=["T-shirt/top", "Trouser", "Pullover", "Dress",
"Coat","Sandal", "Shirt", "Sneaker", "Bag", "Ankle boot"])
```

#output

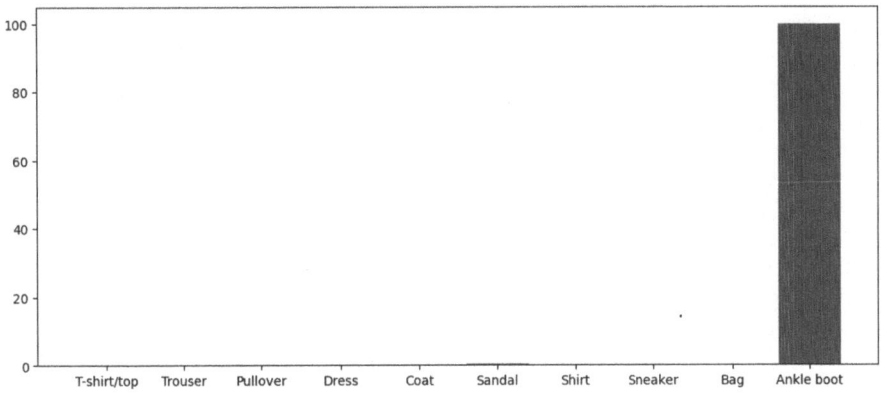

7.5 Summary

This chapter discussed the general architecture of CNNs. CNNs are
useful in computer vision-based applications. The various types of CNNs
are used for specialized purpose to recognize objects. LeNeT, AlexNet,
GoogLeNet, VGGNet, and ResNet were discussed and their detailed
architecture were illustrated. Finally, a case study on computer vison
was implemented in Python with the fashion MNIST data set that clearly
recognized the costumes with an overall test accuracy of 90.14%.

CHAPTER 8

RNNs, LSTMs, and GANs

8.1 Introduction

The previous chapters cover neural networks with light eight neurons that has no additional memory states attached. In this chapter, a new type of neural network called Recurrent neural network (RNN) is focused in which the neurons are attached with memory states to store some information. In addition, the improved RNN type called Long Short-Term Memory (LSTM) and its working principle is planned to discuss. Finally, the Generative Adversarial Networks (GAN) architecture which forms the basic idea for Generative AI is included in this chapter.

8.2 Recurrent Neural Networks

A recurrent neural network (RNN) is a neural network with additional memory state storage attached to each neuron in the network. It is designed in such a way that the memory state holds information about the previous state inputs. Though there exists connectivity between neurons in traditional networks, the input and output are independent of each other, and no retention of any information exists in the neuron. But in an RNN, the retention of information is through a memory state or a hidden state,

© G.R. Kanagachidambaresan and N. Bharathi 2024
G.R. Kanagachidambaresan and N. Bharathi, *Learning Algorithms for Internet of Things,*
Maker Innovations Series, https://doi.org/10.1007/979-8-8688-0530-1_8

which enables the remembrance of previous state information to predict efficiently the next state outcome of sequence-based applications. RNN is the first of its kind that remembers the input using internal memory, which makes it more suitable for sequential data. Unlike traditional neural networks, RNN neurons are dependent of each other and are called *recurrent*, which means they perform sequential computations for applications like time-series forecasting, speech recognition, generating text, etc. RNNs are used in well-known applications like voice search in Google and Apple's Siri.

RNNs have the ability to predict data from the sequential nature of input that other algorithms cannot predict. It is a strong and stable type of neural network, which helps in modeling the sequence data-based problems as it is the only type that has memory in a neuron. It can understand the sequence data and its relation with each other along with the patterns more so than any other neural network-based algorithms. The sequence data must be connected through temporal dynamics rather than its spatial content.

RNN Architecture

An RNN has memory cells in the neurons that play a key role during computation. The memory cell controls the flow of information between input and output layers for keeping track of information and predicting the output. During the forward propagation, it computes the output of each neuron of that layer by updating its internal state in the memory cell and forwards the output to the next layer. During backward propagation through time (BPTT), the partial derivatives of the output is sent back to the network through the sequence of input received. This in turn helps determine the effect of the previous neuron's output to the current cell's output. Then the weights of the network are tuned to get better prediction over the iterations during the training phase.

The formula for finding the current state (hidden) is as follows:

$$h_t = f(h_{t-1}, i_t)$$

where h_t is the current state, h_{t-1} is the previous state, and i_t is the input. After applying the activation function *tanh*, the current state expression is as follows:

$$h_t = tanh(w_{hh}*h_{t-1} + w_{ih}*i_t + b)$$

where w_{hh} is weight of recurrent neuron, w_{ih} is the weight of the input neuron, and b is the bias of hidden layer.

The output is as follows:

$$o_t = w_{ho}h_t + c$$

where w_{ho} is the weight of the output layer and c is the bias at the output layer.

Figure 8-1 illustrates the temporal unfold of an RNN, which depicts the forward feed and backward feed over successive time instances.

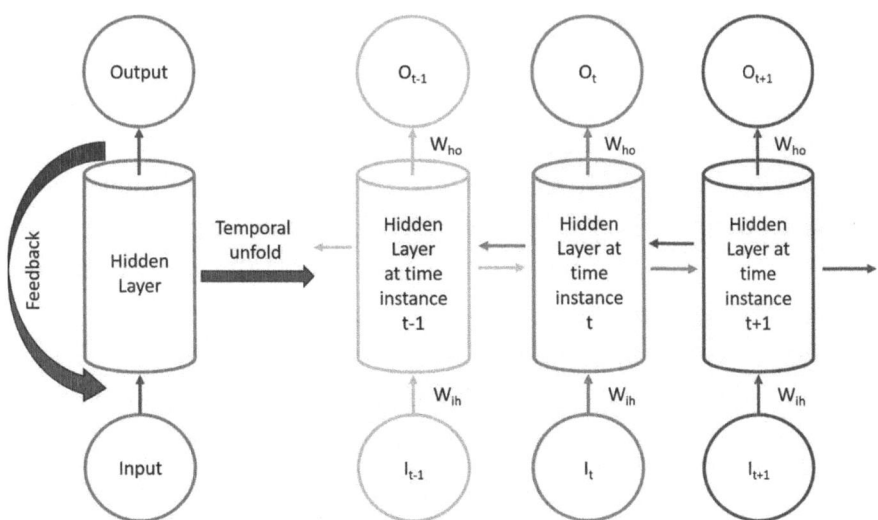

Figure 8-1. *Temporal unfold of RNN*

Types of RNN

The RNN is classified into four types, as illustrated in Figure 8-2, based on the nature or number of inputs and outputs: one-to-one, one-to-many, many-to-one, and many-to-many.

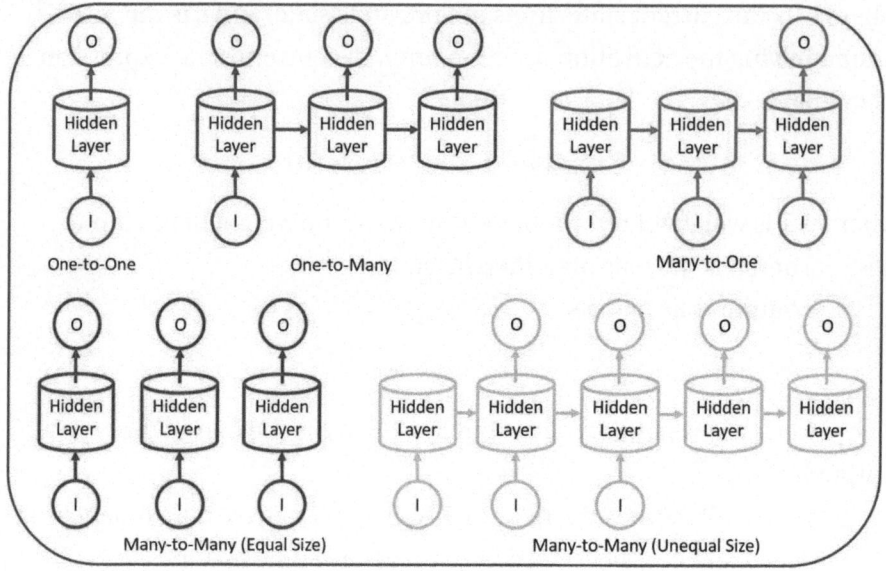

Figure 8-2. *Types of RNN*

One-to-one: This type is the simplest among all types. It has only one input and one output. The input and output sizes are fixed. It is also known as vanilla RNN. This type is recommended for image classification.

One-to-many: This type of RNN generates several outputs from the fixed-size single output. It generates a sequence of outputs that are more suitable for image captioning application. It gives the sentence with multiple words as output from the single fixed-size image input.

Many-to-one: This type of RNN produces the fixed-size single output from the sequence of several inputs. This receives a sentence with a sequence of words as input and generates the classified output stating the sentence is positive or negative or neutral. It is suitable for sentiment analysis applications.

Many-to-many: This type of RNN generates a sequence of several outputs from the sequence of several inputs. This in turn is categorized into two types based on the number of differences between the input and output. If the number of inputs and outputs are the same, then it is called an *equal unit size* and is suitable for name-entity recognition and tagging.

If the number of inputs and outputs are different, then it is called an *unequal unit size* and is quite suitable for language translation.

The RNN is also categorized based on the nature of neural network layer's interconnections that come in the following types:

Vanilla RNN: This is a simple RNN; it is also called an ELMAN network. It has a single layer that produces output from the received input and transfers it to the next time step.

Long short-term memory (LSTM) : Each neuron in LSTM comprises three states: the forget state, which controls which information to retain; the input state, which controls the information extraction from the current input; and the output state, which controls the information to be stored in the next hidden state.

Gated recurrent units (GRUs): This is more like LSTM with only two states: reset and update. It uses a reset gate for short-term memory the which involves updating the hidden state and updating the gate, which updates long-term memory.

Bidirectional RNN: This is a type of RNN that process the input sequences in two directions, combines future and past input contexts to help in prediction. The RNN layer consists of two stacked flows, one moving in the forward direction and the other moving in the reverse direction. The output layer combines the hidden state values to predict the output.

Deep RNN: This is a simple RNN that follows several time steps along the time direction to reach the output from the input. In deep RNNs, in addition to the deep time steps, the depth is increased along the direction of input to output. Hence, each neuron not only passes its hidden state value to the next step but also to the next RNN layer with the same time step along the direction from the input to the output.

Hierarchical RNN: This type handles the hierarchically structured data more efficiently. It comprises multiple stacked layers in which each stack handles a level in the hierarchy. It is suitable for document-based sentiment analysis in which the input data is organized hierarchically as words, sentences, paragraphs, etc.

8.3 Long Short-Term Memory (LSTM)

LSTM is an enhanced version of RNN. RNN stores the output of the previous output for the sake of better future prediction, but it has short-term memory that does not retain the information in the network to learn the long-term dependency. Alternatively, LSTM stores the information with the help of a memory cell in its architecture for a long period and hence supports prediction effectively. RNNs also have downsides such as vanishing and an exploding gradient problem, which is handled efficiently by LSTM. LSTM uses the gates and memory cell to control the flow of information and enables the selective discarding and retaining of data when needed. This helps it to prevent the vanishing gradient and exploding gradient problems. LSTM architectures are robust in learning long-term dependencies from input data, which enables it to work well for speech recognition, time series applications. It can also be combined with convolutional neural networks (CNNs) to get better results in image and video-based analysis.

LSTM Architecture

The LSTM architecture mainly comprises three gates: input gate, forget gate, and output gate, as shown in Figure 8-3. The input gate controls the information stored into the memory cell, the forget cell controls the discarding of information from the memory cell, and finally the output gate controls the reading of information stored in the memory cell. The LSTM network has many such cells connected in succession of one after the other.

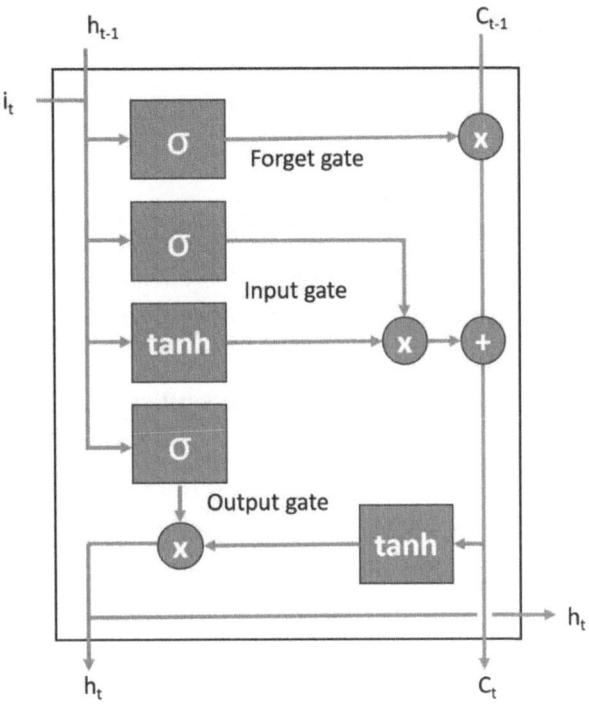

Figure 8-3. *Basic components of LSTM*

The memory cell has two states: the hidden state and the cell state. The states are updated continuously and carry forward the information from one time step to the next. The cell state corresponds to the long-term memory, and the hidden state corresponds to the short-term memory. The hidden state of the current memory cell is updated based on the input, the previous hidden state, and the current cell state. This mechanism allows the LSTM to retain whatever information is necessary and discard the unnecessary information from the memory cell and achieves the learning of long-term dependencies.

A series of steps happens at each cell over the LSTM network to receive the previous input and proceed further to produce the sequence of information. The steps are computing the forget gate, determining the input gate value, updating the cell state based on the previous two steps, and finally computing the hidden state that serves as output to the next cell. Hence, the hidden state carries the information from the previous cell, whereas the cell state gathers all the previous information as the long-term information carrier. This in turn supports LSTM to work well with time-series and sequential applications.

The forget gate is computed using the function $f_t = \sigma\ (W_f\ {}^*[h_{t-1},\ i_t] + b_f)$ where W_f is the weight matrix of forget gate, $[h_{t-1},\ i_t]$ is the concatenation of the previous hidden state and current input, b_f is the bias of the forget gate, and σ is the sigmoid activation function

The input gate value is computed using the function $i_t = \sigma\ (W_i\ {}^*[h_{t-1},\ i_t] + b_i)$ where W_i is the weight matrix of the input gate, and b_i is the bias of input gate.

The cell state is computed using the function $C_t = f_t \odot C_{t-1} + i_t \odot C_t$ where C_{t-1} is the previous state of the cell and $\odot$ is the element-wise operator.

Finally, the hidden state (output) is computed as $o_t = \sigma\ (W_o\ {}^*[h_{t-1},\ i_t] + b_o)$ where W_o is the weight matrix of output gate and b_o is the bias of output gate.

8.4 Bidirectional LSTM Model

BLSTM or BiLSTM is a type of RNN that can process data both in reverse and forward directions. This feature enables the BLSTM to acquire long-term dependencies in the data, which is not supported in simple LSTM as it can process data in the forward direction alone. This type of LSTM has two networks, in which the input information or data flows in two directions: one to process the data in the forward direction and the other one to process the data in a backward direction. Finally, the partial outputs of both directions are combined as sum, multiplication, concatenation, etc., to generate the complete output, as shown in Figure 8-4. It is well suited for the applications that have more connectivity in their input sequence with dependencies among them.

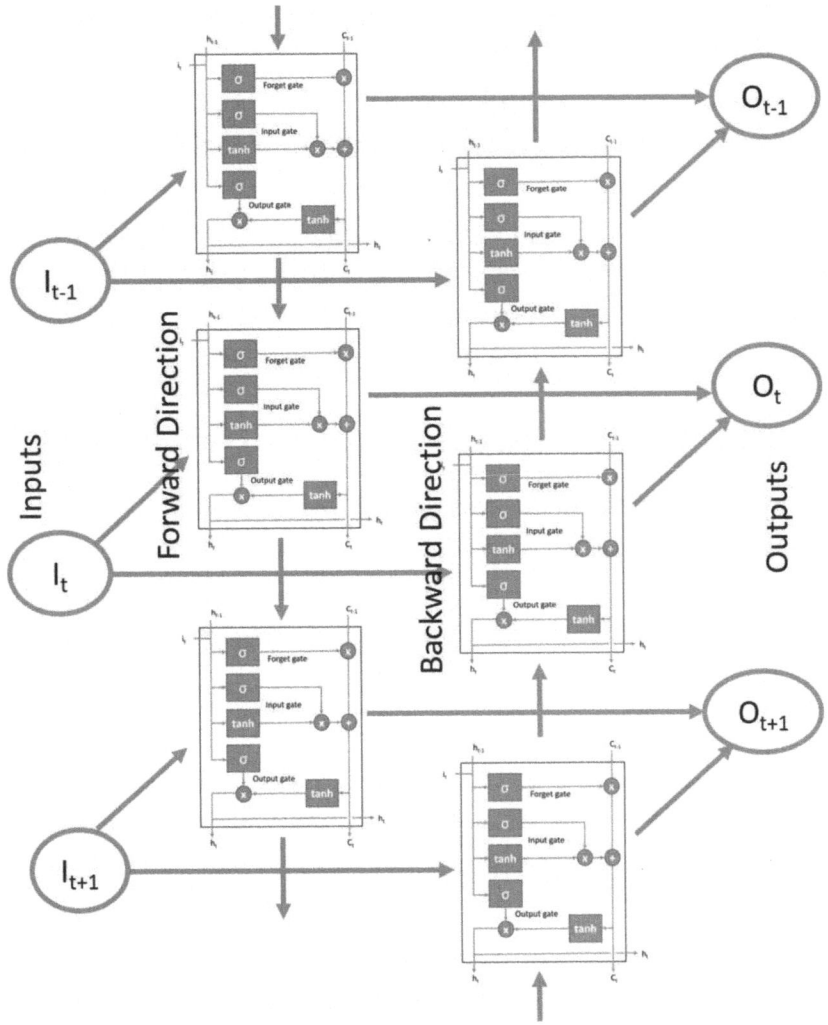

Figure 8-4. *Bidirectional LSTM unfolded architecture*

The applications that use BLSTM are speech recognition, text summarization, etc. It is very much useful in natural language processing applications as any word in the sentence relates to other words in the sentence both in the forward and reverse directions. BLSTM can process the sentence in both the forward and reverse directions and gets the meaningful insights from the input sentence.

For example, "No sentence begins with *because*, because *because* is a conjunction."

In the previous sentence, BLSTM rightly identifies the meaning of the sentence as it processes the sentence in both directions. It is difficult for other models to recognize the correct meaning of this sentence. As it involves the processing of sequenced data in both directions the computation requirement is almost double traditional LSTM, BLSTM should be used only when its real purpose is required.

8.5 Generative Adversarial Networks (GANs)

GANs were a milestone in the field of AI; they generate realistic images, transforming existing styles in real or artificial pictures and generating human faces using "This person does not exist" random face generation framework. It also proves its efficiency by generating images from textual descriptions.

Generator and Discriminator Model

A GAN comprises two phases of neural networks: a generator and a discriminator. The generator is used to synthesize new data, and the discriminator is used to differentiate between a genuine and fake one. Both phases are trained parallelly with adversarial-based training. The generator is fed with noise and produces data very similar to the real data. The discriminator is fed with real data and synthesizes data from generator. It gives the probability of realness in the input. The two phases compete each other to convince the opponent. The generator tries to convince the discriminator in generating fake data. The discriminator in turn tries to determine the fake and informs the authenticity along with a classifier. Obviously, the aim is to produce data that is more realistic fake data by the generator, and the discriminator increases its strength in determining the fake using the efficient training. There are two losses involved, generally during the training phase, which is the generator loss and the discriminator loss. The generator loss depicts the weakness of fake data in showcasing its competency over the real one. The discriminator loss depicts that how far it is believing the fake one as real. The two losses are communicated as depicted in Figure 8-5 through backpropagation to the generator and discriminator, respectively, to improve its process and reduce their corresponding loss.

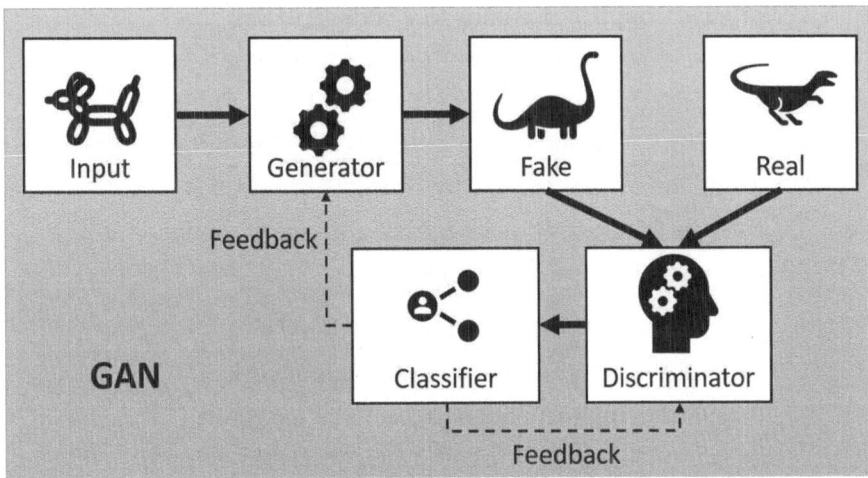

Figure 8-5. *General architecture of GAN*

Types of GAN

The GANs are categorized based on its neural network structure, its process, its suitability for applications, etc. There are various types of GANs, and in this section, we discuss the famous types.

(i) **Vanilla GAN:** This type is simple in architecture with generally two neural networks: generator and discriminator. It is suitable for image and video-based applications but mostly unstable due to its failure to converge at certain times.

(ii)   **Style GAN:** In this type, the generator learns the features of the object it is learning, such as a human face, and applies the learned features to generate the image of a human face that does not exist in the world (e.g., thispersondoesnotexists.com).

(iii) **Disco GAN:** This type is interesting because the objects from two domains (e.g., bike and bag with same color) and how they are related are recognized. Humans can easily identify the relation between two objects from different domains but it is difficult for a learning algorithm. Disco GAN is suitable for generating images of animals with the use of reinforcement learning.

(iv)   **Deep convolutional GAN:** In this type, the generator uses a deep convolutional network to generate high-quality images. ReLu and tanh activation is used in the generator, and leaky-ReLu is used in discriminator to generate and discriminate in better way.

(v)   **Cycle GAN:** In this type, the generator and the discriminator are trained in a cyclic fashion and improves their efficiency. This type is well suitable to transform an image into another form such as a horse to a zebra, winter to summer, a Monet painting to a photo, and vice versa.

(vi)   **Conditional GAN:** In this type, the generator is fed with more information such as labels or values of features other than the generator loss through back propagation. The positive side of this type is that the generator can be tuned to generate images the same as the training images.

(vii) **Least square GAN:** The loss function follows the least square method for the discriminator loss computation. In this type, if the objective function using the least square method is minimized, then it leads to the minimization of divergence and converge faster.

(viii) **Text to image synthesis:** Generally GANs are generating images randomly after learning the features of the object; instead, an object that is constructed from the description of text will be more natural and very much appropriate for business visuals and use cases.

(ix) **Auxiliary Classifier GAN:** This is an enhanced version of CGAN in which the image is classified easily using labels; besides, the source of the classified image can be determined.

(x) **Dual Video Discriminator GAN:** This type mainly focuses on videos, and the video generation is based on the bigGAN framework. Particularly in this type the discriminator handles the videos with spatial and temporal discrimination.

(xi) **Super Resolution GAN:** SRGAN is used to transform the low-resolution images into higher resolution. It can produce high-resolution images without any additional information and can potentially compete with real high-resolution images.

(xii) **Stack GAN:** This type contains stacked generators and discriminators connected in series so that each feature is handled by each generator in the series and the same-positioned discriminator uses the same feature to discriminate the generated image.

(xiii) **Info GAN:** This type of GAN is used to generate not only realistic images in nature but also more comprehensive and meaningful images when viewed. This GAN uses a more efficient loss function that helps to generate informative, realistic images.

(xiv) **Pixel RNN:** This uses explicit data distribution to predict the next pixel in the image. It uses an RNN to determine the features per pixel using 1 x 1 convolution.

8.6 Application Case Study

Generally, one form of information is converted into another form to provide a better understanding for the users of applications. Text-to-speech and Speech-to-text conversion is required for reducing strain on the eyes, driving assistance, automatic customer calls, subtitles and transcriptions, low reading or hearing ability people, etc.

Text to Speech

This module requires the installation of a few language-related libraries that are used to convert text to speech. The library pyttsx3 is a Python text-to-speech X platform that initializes a speech engine through the engine interface. Another library called espeak is a speech synthesizer library that generates clear, faster English and a few other languages. gTTS is the Google Translates Text to Speech library, which makes it easier to convert text to speech using Python script.

```
pip install pyttsx3
!apt-get install espeak
pip install gTTS
```

```
# permission for accessing the drive folder as this program is
trying to access the text file
# using google colab
from google.colab import drive
drive.mount('/content/gdrive')

# linking the library pyttsx3 and initializing the
speech engine
import pyttsx3
sp_eng = pyttsx3.init()
sp_eng.runAndWait()

# importing the operating system interface to change the
directory where the text file is saved
# and the content is read into a string object.
import os
os.chdir("/content/gdrive/My Drive/Colab Notebooks/LAIoT/
Chapter 8")
print(os.getcwd())
print(os.listdir())
fptr = open("content.txt", "r")
sentence = fptr.read()

#gTTS converts the text input into a audio object which can be
stored as .mp3 file
from gtts import gTTS
speak_lang = 'en'
top_domain = 'co.in'
speech = gTTS(text=sentence, lang=speak_lang, tld = top_domain,
slow=False)
speech.save("firstspeech.mp3")

# The output file firstspeech.mp3 is saved in the current
directory of the google drive
```

Speech to Text

This module converts the speech into text by using a speech recognition library. It uses the Google Speech API. This library needs to be installed before calling the recognizer and recognize Google methods. This module also uses the pydub library in Python to manipulate audio files such as play, edit, merge/split, etc. The only constraint in using the pydub library is that it manipulates only .wav files.

```
!pip install SpeechRecognition
!pip install pydub

# permission for accessing the drive folder as this program is
trying to access the speech file
# using google colab

from google.colab import drive
drive.mount('/content/gdrive')

# importing the operating system interface to change the
directory where the speech file is
# saved and the content is read into a string object.
import os
os.chdir("/content/gdrive/My Drive/Colab Notebooks/LAIoT/
Chapter 8")

#audio file with different extension could be extracted and
exported as .wav file
from pydub import AudioSegment
sd_conv = AudioSegment.from_mp3("firstspeech.mp3")
sd_conv.export("firstspeech.wav", format="wav")
print(sd_conv)

# speech recognition library is linked and the recognizer is
called in order to convert the
# audio .wav file into corresponding text file
```

```
import speech_recognition as sp_rg
sr = sp_rg.Recognizer()

with sp_rg.AudioFile('firstspeech.wav') as sp_file:
    aud_txtobj= sr.listen(sp_file)
    print(aud_txtobj)
```

```
# The audio file is extracted and displayed as text and
otherwise flags exception.
    try:
        txt_data = sr.recognize_google(aud_txtobj)
        print("The speech file contains information as
        follows: ")
        print(txt_data)
    except:
        print('Exception occured...')
```

The Output as

```
Hi hello How are you I am fine
```

8.7 Summary

The chapter explained the basic recurrent neural network and its architecture in detail. The various types of RNNs were demonstrated using figures. Besides the types, neural network connectivity was discussed. Then the most efficient type of RNN, the LSTM, was covered, with its basic architecture, and the mathematical expressions for all three gates were discussed. Then the powerful bidirectional LSTM was explained so that its efficiency could be fully utilized. Finally, the adversarial generator and discriminator model that frames GAN was described along with its pros and cons.

CHAPTER 9

Optimization Methods

9.1 Introduction

Learning algorithms have several parameters such as learning rate, epochs, etc. The output of a learning algorithm is compared to the actual output, and the difference between them is calculated and called the *cost function*. Optimizers are functions that are used in machine learning algorithms to minimize the difference between the actual output and the predicted output using a gradient. The minimization of the cost function value is carried out by finding the gradient. The gradient is the value change of all the parameters involved in learning with respect to the change happening in the cost function. The slope can be steeper if the gradient is higher, and learning happens in a faster way. This will repeat for several iterations in order to minimize the cost function value. But at the same time, the learning process stops if the slope becomes zero.

Types of Optimizers

The optimizers update the weights involved in the learning algorithm along with the learning rate to move toward the minimum cost function value. It will repeat for certain iterations, and it stops at a point where further minimization is negligible with a corresponding computing time. It updates the weight as follows:

$$\left\{ W_x = W_{x-1} - \alpha \frac{\partial \theta}{\partial W_x} \right\}$$

© G.R. Kanagachidambaresan and N. Bharathi 2024
G.R. Kanagachidambaresan and N. Bharathi, *Learning Algorithms for Internet of Things,*
Maker Innovations Series, https://doi.org/10.1007/979-8-8688-0530-1_9

where α is the learning rate, θ is the parameter involved in learning, and W_x is the weight that is updated in every iteration during optimization.

The various types of optimizers based on the method of applying the gradient and additional parameters like momentum, etc., are as follows:

- Gradient descent (GD) optimizer

- Batch gradient descent optimizer

- Stochastic gradient descent optimizer

- Mini-batch gradient descent optimizer

- Adagrad

- RMSProp

- Adadelta

- Momentum

- Nesterov momentum

- Adam

- Adamax

- SMORMS3

9.2 Gradient Descent

This type is a commonly used type of learning algorithms and also the oldest one. It is suitable for convex functions where a unique global minimum exists. Updating weights over iterations leads to fixing the best values to the parameters involved in learning and results in a global minimum. But the challenge is that this type is not suitable for nonconvex functions, as it may end in one of the local minima instead of the global minima. Since nonconvex functions have many peaks and valleys, multiple local minima exists, and this type may be bounded in any one of the local minima instead of the global minima.

Another challenge is a saddle point problem where the gradient becomes zero and is not an optimal value. Besides, very small and very large gradient values lead to vanishing gradient or exploding gradient problems. This is mainly because of the incorrect initialization of learning parameters, and it results in high-cost function values due to nonconvergence. The following code and Figure 9-1 demonstrate this optimizer:

```
import matplotlib.pyplot as mplt
import numpy as npy
from mpl_toolkits.mplot3d import Axes3D

# function to be optimized
def f(x, y):
    return x**2 + y**2

# partial derivatives of the function based on x and y
def df_wrt_dx(x, y):
    return 2 * x

def df_wrt_dy(x, y):
    return 2 * y

# Gradient descent function
def G_D(xyval, l_rate, num_itrns):
    # Initialize x and y
    x = xyval[:, 0] + npy.random.rand(len(xyval)) *
    (xyval[:, 1] - xyval[:, 0])
    y = xyval[:, 0] + npy.random.rand(len(xyval)) *
    (xyval[:, 1] - xyval[:, 0])
    series = []
```

```
    # Perform the gradient descent iterations
    for i in range(num_itrns):
        # gradients computation
        grad_x = df_wrt_dx(x, y)
        grad_y = df_wrt_dy(x, y)

        # parameters updation
        x = x - l_rate * grad_x
        y = y - l_rate * grad_y

        # Save the parameter series
        series.append((x[0], y[0], f(x[0], y[0])))

    return series

# input specification

xyval = npy.asarray([[-1.0, 1.0], [-1.0, 1.0]])
x_val = npy.linspace(xyval[0, 0], xyval[0, 1], 100)
y_val = npy.linspace(xyval[1, 0], xyval[1, 1], 100)
X, Y = npy.meshgrid(x_val, y_val)
Z = f(X,Y)

# gradient descent algorithm execution
l_rate = 0.1
num_itrns = 20
series = G_D(xyval, l_rate, num_itrns)

# plotting of results in 3D plot
fig1 = mplt.figure()
graf = fig1.add_subplot(111, projection='3d')
graf.plot_surface(X, Y, Z, cmap='viridis', alpha = 0.5)
graf.scatter(*zip(*series), c='r', marker='o')
graf.scatter(series[0][0], series[0][1], series[0][2],
    color='red', label='Best_solution')
```

```
x_series = [point[0] for point in series]
y_series = [point[1] for point in series]
z_series = [point[2] for point in series]

graf.plot(x_series, y_series, z_series,
    color='blue', label='Trajectory from intitial point')
graf.set_xlabel('x')
graf.set_ylabel('y')
graf.legend()
mplt.show()
```

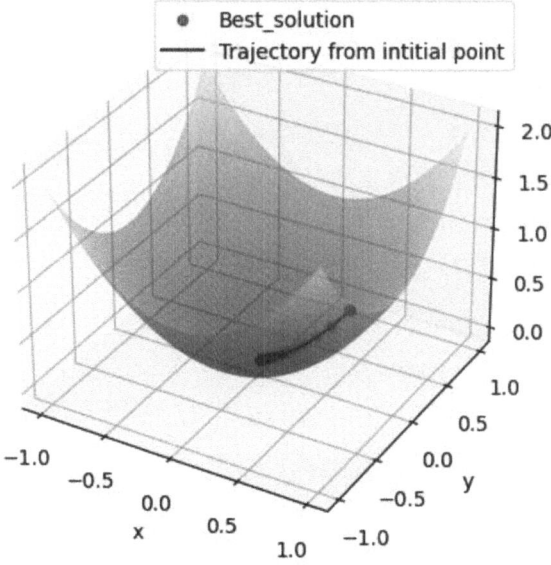

Figure 9-1. *Demonstration of gradient descent optimizer*

9.3 Batch Gradient Descent

This type is a variant of gradient descent that involves a complete training set during each iteration, and hence it is more accurate. It takes the average of the gradients of all training data samples, computes the mean gradient, and updates the parameters using the mean gradient. The cost function value decreases smoothly as it progresses over the epochs. This type is well suited with the increased number of features. Though it generates accurate output and comparatively smooth error manifolds, the computation cost is very high.

The downside of this optimizer is that it takes relatively more time for convergence as it takes the entire training set during each iteration. This type is not good for the nonconvex functions as it handles the entire set of input in each iteration; therefore, there exists the possibility of getting trapped in local minima. The following code and Figure 9-2 demonstrate the convergence of this optimizer:

```
#definition of linear function
def lin_func(X, t):
 # assert X.ndim > 1
 # assert t.ndim > 1
  if(X.ndim >1) and (t.ndim > 1):
    return npy.dot(X, t)

#Mean Square Loss function(MSE)
def MSE_fn(yactual, ypred):
  return npy.mean((yactual - ypred)**2)

#Initialize the weight parameter
def init_tta(W):
  return npy.zeros([W, 1])

#computation of gradient using the derivative with respect to
the weight
def B_grad(X, y, tta):
  return -2.0 * npy.dot(X.T, (y - lin_func(X, tta)))

#gradient descent updation
def up_func(tta, grad, stp_sz):
  return tta - stp_sz * grad
```

```
def BGD_train(X_train, y, epochs, stp_sz=0.1, plt_all=1):
  N, D = X.shape
  tta = init_tta(D)
  loss_val = []
  for epoch in range(epochs):
    ypred = lin_func(X, tta)
    loss = MSE_fn(y, ypred)
    grads = B_grad(X, y, tta)
    tta = up_func(tta, grads, stp_sz) # Updation of parameters
    using the gradients
    loss_val.append(loss)
    print(f"\nEpoch:{epoch}, Mean Square Error: {loss}")
  return loss_val

import numpy as npy

X = npy.array([[0.88837122, 0.55088876, 0.07233497, 0.18170872,
0.75790715],
[0.99225237, 0.5395178, 0.79842092, 0.02775931, 0.39621768],
[0.30150944, 0.68509148, 0.3827384, 0.83898851, 0.7266716]])

y = npy.array([[0.67738928], [0.81657103], [0.13115408]])

loss_val = BGD_train(X, y, 20, stp_sz=0.1, plt_all=1)

import matplotlib.pyplot as mplt
mplt.plot(loss_val)
mplt.xlabel("Training iteration / Epoch")
mplt.ylabel("loss_val")
mplt.title("BGD learning curve")
```

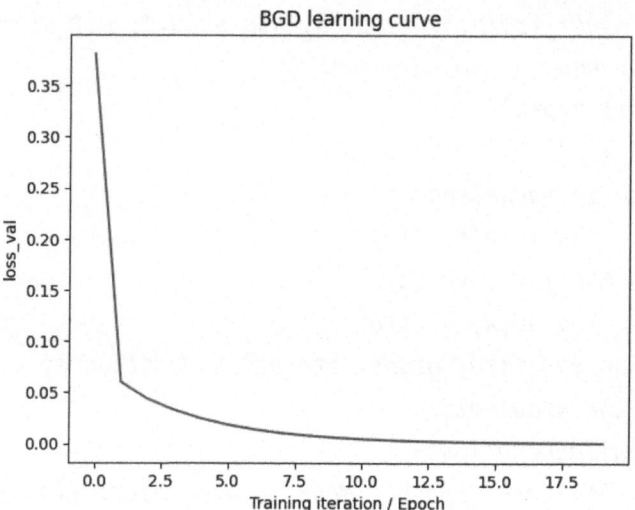

Figure 9-2. *Demonstration of convergence of batch gradient descent*

9.4 Stochastic Gradient Descent

This type is also a variant of the gradient descent that can be used for nonconvex functions. Generally, the nonconvex problems have many local minima, which may mislead the optimizer to end up with local minima instead of the global minima. In this type, the mistake is overcome by batch processing, concentrating on one update at a time. Hence, it updates the weights in a faster way with high variance and fluctuations of cost function values. This is also the reason for reaching the global minimum sooner.

The downside of this type is that the convergence happens in a slower way if the learning rate is too small and swings around the minimum or deviates from the global minimum if the learning rate is very large. The following code and Figure 9-3 illustrate this optimizer. Refer to the previous code section for certain functions.

```
def randshuffle_data(X, y):
  N, _ = X.shape
  rand_idx = npy.random.permutation(N)
  return X[rand_idx], y[rand_idx]

def Each_sample_grad(xi, yi, tta):
  return -2.0 * xi * (yi - lin_func(xi, tta))

#training using stochastic Gradient descent
def SGD_train(X, y, epochs, stp_sz, plt_all=1):
  N, D = X.shape
  tta = init_tta(D)
  loss_val = []
  epoch = 0
  loss_tol_limit = 0.001
  mean_loss = float("inf")

  while epoch < epochs and mean_loss > loss_tol_limit:
    current_loss = 0.0
    rand_x, rand_y = randshuffle_data(X, y)

    for ind in range(rand_x.shape[0]):
      select_x = rand_x[ind].reshape(-1, D)
      select_y = rand_y[ind].reshape(-1, 1)
      ypred = lin_func(select_x, tta)
      loss = MSE_fn(select_y, ypred)
      current_loss += loss
      grads = Each_sample_grad(select_x, select_y, tta)
      tta = up_func(tta, grads, stp_sz)

    mean_loss = current_loss/ X.shape[0]
    loss_val.append(mean_loss)
    print(f"\nEpoch:{epoch}, Mean Square Error: {loss}")
    epoch += 1
```

```
    return loss_val

import numpy as npy

#input parameters
X = npy.array([[0.88837122, 0.55088876, 0.07233497, 0.18170872,
0.75790715],
[0.99225237, 0.5395178, 0.79842092, 0.02775931, 0.39621768],
[0.30150944, 0.68509148, 0.3827384, 0.83898851, 0.7266716]])
y = npy.array([[0.67738928], [0.81657103], [0.13115408]])
loss_val = SGD_train(X, y, 20, stp_sz=0.1, plt_all=1)

import matplotlib.pyplot as mplt
mplt.plot(loss_val)
mplt.xlabel("Training iteration / Epoch")
mplt.ylabel("loss_val")
mplt.title("SGD learning curve")
```

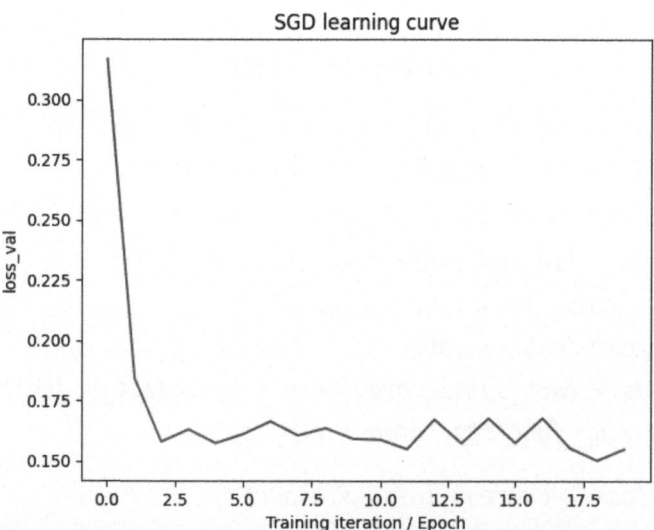

Figure 9-3. *Demonstration of convergence of stochastic gradient descent*

9.5 Mini-Batch Gradient Descent

This type is an amalgamation of batch gradient descent and stochastic gradient descent. Here we neither use entire training data nor use a single data point at a time. Instead, a small batch of training data points are used, called a *mini-batch* as its size is smaller than the entire dataset. Hence, a mini-batch gradient descent updates the parameters involved in learning frequently along with a faster computation time and less storage comparatively. It offers smooth convergence than stochastic and less time consumption than batch gradient descent. Since the entire training set is divided into mini-batches, this type is more suitable for harnessing the parallel computing capabilities by facilitating simultaneous calculations and updating parameters by combining the results. The batches are chosen at random, ranging from 50 to 250 samples.

The downside of this optimizer is that the decision about batch size is still a challenging one, and a lot of experimentation is required to find the optimum batch size. Also, the learning rate with a high value leads to diverging the optimization and with a low value leads to slow convergence. The following code and Figure 9-4 illustrate this optimizer. Refer to the previous code section for certain functions.

```
def MBGD_train(X, y, epochs, stp_sz=0.1, bch_sz=3, plt_all=1):
  N, D = X.shape
  tta = init_tta(D)
  loss_val = []
  num_bch = N//bch_sz
  X, y = randshuffle_data(X, y)

  for epoch in range(epochs):
    current_loss = 0.0

    for bch_ind in range(0, N, bch_sz):
```

```
        x_bch = X[bch_ind: bch_ind + bch_sz] # operate on a batch
        of features
        y_bch = y[bch_ind: bch_ind + bch_sz]

        ypred = lin_func(x_bch, tta) # predict current parameters
        loss = MSE_fn(y_bch, ypred) # MSE calcuation
        grads = B_grad(x_bch, y_bch, tta) # gradients calculation
        tta = up_func(tta, grads, stp_sz) # parameter updation
        current_loss += (loss * x_bch.shape[0]) # batch loss
        calculation

    mean_loss = current_loss/ N
    loss_val.append(mean_loss)
    print(f"\nEpoch:{epoch}, Mean Square Error: {loss}")
  return loss_val

import numpy as npy
#input parameters
X = npy.array([[0.88837122, 0.55088876, 0.07233497, 0.18170872,
0.75790715],
[0.99225237, 0.5395178, 0.79842092, 0.02775931, 0.39621768],
[0.30150944, 0.68509148, 0.3827384, 0.83898851, 0.7266716]])
y = npy.array([[0.67738928], [0.81657103], [0.13115408]])

loss_val = MBGD_train(X, y, 20, stp_sz=0.1, plt_all=1)
import matplotlib.pyplot as mplt
mplt.plot(loss_val)
mplt.xlabel("training Iteration/Epoch")
mplt.ylabel("loss_val")
mplt.title("MBGD learning curve")
```

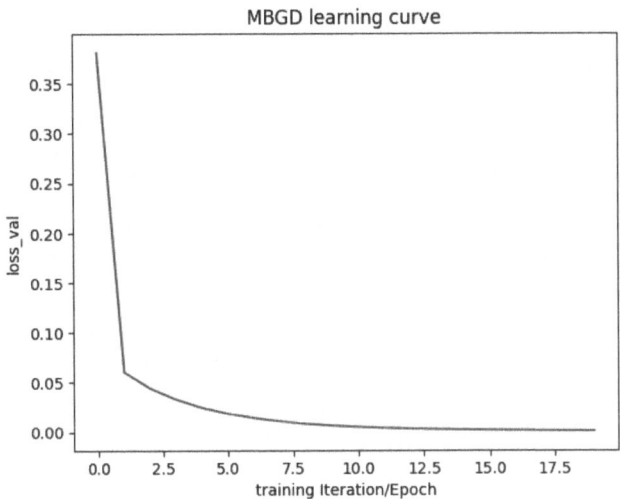

Figure 9-4. *Demonstration of convergence of mini-batch gradient descent*

9.6 Adagrad

This type is said to be an *adaptive* gradient because the learning rate changes for each iteration. In this type, the learning rate plays a major role in updating the parameters involved in learning. This is mainly because of the different frequency of features influencing the learning. The weights of high-frequency features are updated with low learning rates, and the weights of low-frequency features are updated with high learning rates to get better accuracy. This type of optimizer is best suited for sparse data. Hence, at each iteration, the learning rate α is calculated based on the time instance t and other parameters like frequency of features, etc. Also, in this type, there is no need to change the learning rate manually.

The downside of the optimizer is that as the number of iterations increases, the α value becomes very large and leads to the vanishing gradient problem where there is no change between the previous and current iteration cost function value. The following code and Figure 9-5 demonstrate this optimizer:

```
import matplotlib.pyplot as mplt
import numpy as npy
import math
from mpl_toolkits.mplot3d import Axes3D

# function to be optimized
def f(x, y):
  return x**2 + y**2

# partial derivatives of the function based on x and y
def df_dx_dy(x, y):
  return npy.asarray([2.0 * x, 2.0 * y])

# Adagrad function
def adaG(f, df_dx_dy, xyval, num_itrns, stp_sz):
  # record solutions
  solns = list()
  sol_scores = []
  sol_traj = []
  # setting initial position
  soln = xyval[:, 0] + npy.random.rand(len(xyval)) * (xyval[:,
  1] - xyval[:, 0])
  # compute sum of square gradients
  sq_gd_sums = [0.0 for _ in range(xyval.shape[0])]
  # gradient descent calculation
  for itr in range(num_itrns):
    grad = df_dx_dy(soln[0], soln[1])
    # Calcualte squared partial derivatives and add
    for i in range(grad.shape[0]):
      sq_gd_sums[i] += grad[i]**2.0
    # record new solution
    new_soln = list()
    for i in range(soln.shape[0]):
```

```
    # learning rate calculation
    alfa = stp_sz / (1e-8 + math.sqrt(sq_gd_sums[i]))
    # new position calculation
    pos = soln[i] - alfa * grad[i]
    new_soln.append(pos)
  # Record current solution
  soln = npy.asarray(new_soln)
  solns.append(soln)
  # Evaluate the solution
  soln_eval = f(soln[0], soln[1])
  sol_score = f(soln[0], soln[1])
  sol_scores.append(sol_score)
  sol_traj.append(soln.copy())

 return soln, sol_scores, sol_traj
# Set the seed of random number generator
npy.random.seed(4)
# set the input
xyval = npy.asarray([[-1.0, 1.0], [-1.0, 1.0]])
# set no. of iterations
n_itr = 50
# set step size
stp_sz = 0.1
# call adagrad function
solns,sol_scores, sol_traj = adaG(f, df_dx_dy, xyval,
n_itr, stp_sz)

x = npy.linspace(xyval[0, 0], xyval[0, 1], 100)
y = npy.linspace(xyval[1, 0], xyval[1, 1], 100)
X, Y = npy.meshgrid(x, y)
Z = f(X, Y)
```

```
solns = npy.asarray(solns)
# Plot trajectory path
figure = mplt.figure()
grph = figure.add_subplot(111, projection='3d')
grph.plot_surface(X, Y, Z, cmap='viridis', alpha=0.5)
grph.scatter(solns[0], solns[1], f(solns[0], solns[1]),
    color='red', label='Best_solution')
grph.plot([pt[0] for pt in sol_traj],
    [pt[1] for pt in sol_traj], sol_scores,
    color='blue', label='Trajectory from intitial point')
grph.set_xlabel('X')
grph.set_ylabel('Y')
grph.legend()

# Show the plot
mplt.show()
```

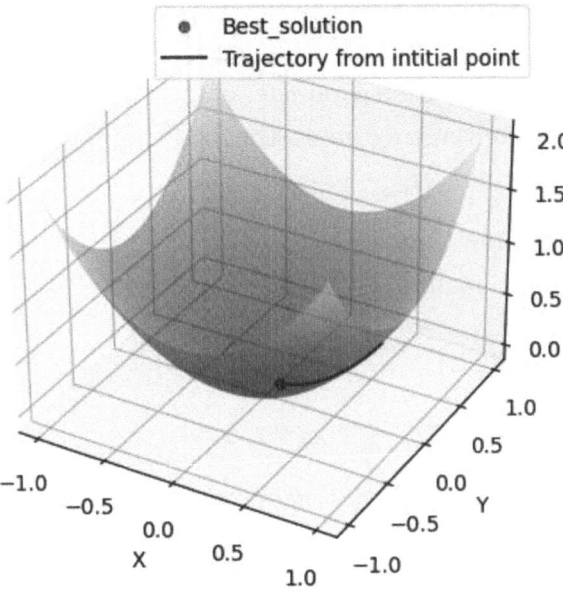

Figure 9-5. *Demonstration of adagrad*

9.7 RMSProp

The root mean square propagation optimizer is an extension of adagrad. This type overcomes the limitations of adagrad that has a monotonically decreasing learning rate. It computes the moving average of squares of the gradient for individual weights. The learning rate is divided by this moving average to ensure that the learning rate for each weight adapts. This type uses the moving average to discard the past rate with exponential decay. This helps the optimizer to reach the converging point sooner after discovering the convex portion of the function used.

The gradient is calculated with $g_t = \delta C / \delta w$, where C is the cost function. Squared gradients are calculated using $E[grad^2]_t = \beta E[grad^2]_{t-1} + (1-\beta) grad_t^2$, where β is the decline rate, generally set to 0.9. Now the adaptive learning rate is computed using $\eta_t = \eta / \sqrt{(E[grad^2]_t + \varepsilon)}$, where η is the initial learning rate and ε is a very less valued constant to avoid division by zero, often set to 1e^{-8}. Update the weight by $W_{t+1} = W_t - \eta_t * grad_t$. This process is carried out for each parameter of the network for epochs count or until the convergence point. The following code and Figure 9-6 demonstrate this optimizer:

```
import numpy as npy
import matplotlib.pyplot as mplt

# function to be optimized
def f(x1, x2):
    return 7 * x1**2.0 + 5 * x2**2.0

# partial derivatives of the function based on x1 and X2
def df_x1(x1, x2):
    return 14.0 * x1

def df_x2(x1, x2):
    return 10.0 * x2

def rmsprop(x1, x2, df_x1, df_x2, lr_rate, epsilon, gamma,
epochs_limit):
    # store x1, x2 and y trajectories
    x1_traj = []
    x2_traj = []
    y_traj = []

    # Set values for x1, x2, and y
    x1_traj.append(x1)
    x2_traj.append(x2)
    y_traj.append(f(x1, x2))

    # initialize e1 and e2
    e1 = 0
    e2 = 0
```

```
    # gradient descent computation
    for _ in range(epochs_limit):
        # compute partial derivatives of function w.r.t
        x1 and x2
        gradt_x1 = df_x1(x1, x2)
        gradt_x2 = df_x2(x1, x2)

        # computing the exponential weighted averages of the
        partial derivatives
        e1 = gamma * e1 + (1 - gamma) * gradt_x1**2.0
        e2 = gamma * e2 + (1 - gamma) * gradt_x2**2.0

        # update x1 and x2
        x1 = x1 - lr_rate * gradt_x1 / (npy.sqrt(e1 + epsilon))
        x2 = x2 - lr_rate * gradt_x2 / (npy.sqrt(e2 + epsilon))

        # updated values of x1, x2 and y are appended in
        its list
        x1_traj.append(x1)
        x2_traj.append(x2)
        y_traj.append(f(x1, x2))

    return x1_traj, x2_traj, y_traj

# initialization of parameters
init_x1 = -4.0
init_x2 = 3.0
lr_rate = 0.1
gamma = 0.9
epsilon = 1e-8
epochs_limit = 50

# RMSprop execution
```

```
x1_traj, x2_traj, y_traj = rmsprop(init_x1,init_x2,df_x1,df_
x2,lr_rate,epsilon,gamma,epochs_limit)

# plot in 3D
x1 = npy.arange(-5.0, 5.0, 0.1)
x2 = npy.arange(-5.0, 5.0, 0.1)

# Creating a meshgrid of x1 and x2
x1, x2 = npy.meshgrid(x1, x2)

# Computing y component
y = f(x1, x2)

fig1 = mplt.figure()
graph3d = fig1.add_subplot(111, projection='3d')
graph3d.plot_surface(x1, x2, y, cmap='viridis', alpha=0.5)
graph3d.scatter(x1_traj[0], x2_traj[0], f(x1_traj[0],
x2_traj[0]), color='red', label='Best solution')
graph3d.plot(x1_traj, x2_traj, y_traj, color='blue',
label='Trajectory from initial point')

graph3d.set_xlabel('x1')
graph3d.set_ylabel('x2')
graph3d.set_zlabel('y')
graph3d.legend()
graph3d.set_title('RMSProp learning')
```

RMSProp learning

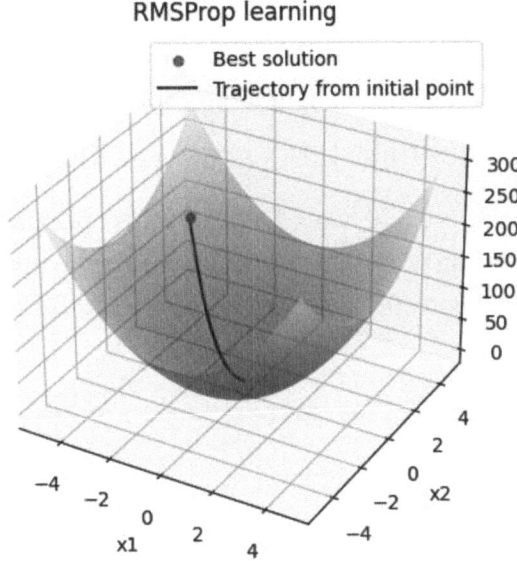

Figure 9-6. Demonstration of RMSProp

9.8 Adadelta

This is also an extension of adagrad. It uses the sum of gradients as a weighted average of square of fixed number of past gradients instead of using the entire history of gradients. This in turn maintains the learning rate not to reduce to a very small value. The formula for weight updating is the same, but the computation of the learning rate at each iteration is with a weighted average of a fixed number of past gradients.

The moving average $E[grad^2]_t$ at time instant t is given by $E[grad^2]_t = \gamma E[grad^2]_{t-1} + (1-\gamma)grad_t^2$.

Where γ is the learning rate generally set to 0.9.

Let the parameter vector update be $\Delta\theta_t = -\eta * grad_{t,i}$ and $\theta_{t+1} = \theta_t + \Delta\theta_t$

The vector of Adagrad is $\Delta\theta_t = (-\eta / \sqrt{(G_t + \epsilon)}) * grad_t$, and it is rewritten using average past squared gradients as $\Delta\theta_t = (-\eta / \sqrt{(E[grad^2]_t + \epsilon)}) * grad_t$.

This can then be written as $\Delta\theta_t = (-\eta /RMS[grad]_t))* grad_t$. Now the learning rate η is replaced with the updated RMS parameter until the previous time step $RMS[\Delta\theta]_{t-1}$. Then the $\Delta\theta_t$ is as follows:

$$\Delta\theta_t = (-RMS[\Delta\theta]_{t-1} /RMS[grad]_t))* g_t$$

The updated rule is $\theta_{t+1}=\theta_t + \Delta\theta_t$. Hence, in this optimizer, there is no need to set the learning rate as it is eliminated from the updated rule. The following code and Figure 9-7 demonstrate this optimizer:

```
import numpy as npy
import matplotlib.pyplot as mplt
import math

# function to be optimized
def f(x, y):
 return 7.0*x**2.0 + 5.0*y**2.0

# partial derivatives of the function based on x and y
def df_dx_dy(x, y):
 return npy.asarray([x * 14.0, y * 10.0])

# Adadelta implementation
def AD(f, df_dx_dy, xyval, num_itr, rhow, ep=1e-3):
 # record all solutions
 solns = list()
 sol_scores = []
 sol_traj = []
 # set an initial position
 soln = xyval[:, 0] + npy.random.rand(len(xyval)) * (xyval
 [:, 1] - xyval[:, 0])
 # initialize list for storing average square gradients
 mean_sqr_grad = [0.0 for _ in range(xyval.shape[0])]
 # list of the average parameter updates
 mean_sqr_param = [0.0 for _ in range(xyval.shape[0])]
```

```
# computation of gradient descent
for itr in range(num_itr):
 grad = df_dx_dy(soln[0], soln[1])

 for i in range(grad.shape[0]):
   # compute squared gradient
   sqgrad = grad[i]**2.0
   # compute squared gradient moving average and update
   mean_sqr_grad[i] = (mean_sqr_grad[i] * rhow) + (sqgrad *
   (1.0-rhow))
   # record new solution
 new_soln = list()
 for i in range(soln.shape[0]):
    # compute the step size of this variable
    alfa = (ep + math.sqrt(mean_sqr_param[i])) / (ep + math.
    sqrt(mean_sqr_grad[i]))
    # record the variation
    variation = alfa * grad[i]
    # compute squared parameter moving average and update
    mean_sqr_param[i] = (mean_sqr_param[i] * rhow) +
    (variation**2.0 * (1.0-rhow))
     # Compute the next position using this variable and
     record it
    value = soln[i] - variation
    new_soln.append(value)
 soln = npy.asarray(new_soln)
 solns.append(soln)
 soln_eval = f(soln[0], soln[1])
 # display progress
 print(f"\nIteration: {itr}, solution :{soln}, evaluation of
 solution: {soln_eval}")
 # append solution and trajectory of this iteration
```

```
    sol_score = f(soln[0], soln[1])
    sol_scores.append(sol_score)
    sol_traj.append(soln.copy())
  return soln, sol_scores, sol_traj

# set the seed value for random number generator
npy.random.seed(1)
# set limit for input
xyval = npy.asarray([[-1.0, 1.0], [-1.0, 1.0]])
# set the number of iterations
num_itr = 200
# set the rhow value
rhow = 0.99
# call adadelta function
solns, sol_scores, sol_traj = AD(f, df_dx_dy, xyval,
num_itr, rhow)

#plot the 3D visualization of trajectory
x = npy.arange(xyval[0, 0], xyval[0, 1], 0.1)
y = npy.arange(xyval[1, 0], xyval[1, 1], 0.1)
X, Y = npy.meshgrid(x, y)
Z = f(X, Y)

solns = npy.asarray(solns)
# Plot the trajectory
fig1 = mplt.figure()
grph = fig1.add_subplot(111, projection='3d')
grph.plot_surface(X, Y, Z, cmap='viridis', alpha=0.5)
grph.scatter(solns[0], solns[1], f(solns[0], solns[1]),
    color='red', label='Best solution')
grph.plot([point[0] for point in sol_traj],
    [point[1] for point in sol_traj], sol_scores,
    color='blue', label='Trajectory from initial point')
```

```
grph.set_xlabel('X')
grph.set_ylabel('Y')
grph.legend()

# visualize the plot
mplt.show()
```

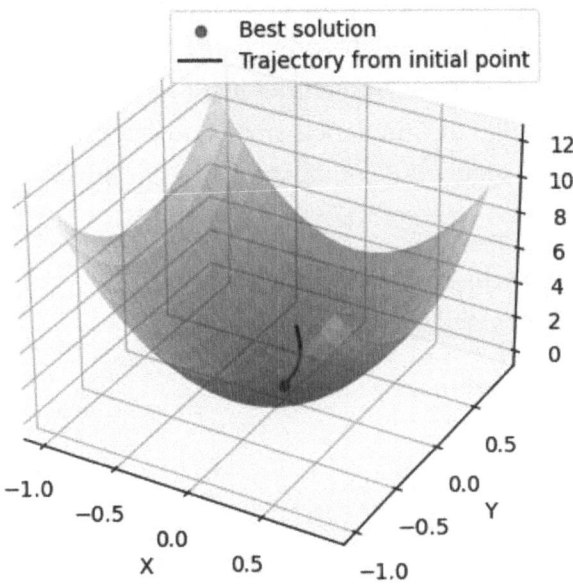

Figure 9-7. *Demonstration of RMSProp*

9.9 Momentum

This type overcomes the downside of all variants of gradient descent by adding momentum as the exponentially weighted average of a fixed number of previous gradients.

This is an extension of stochastic gradient and is best for nonconvex functions. It reduces the number of oscillations in the traditional gradient descent optimizers by accelerating the process to find the global

275

minimum. Momentum is generally set to a value between 0 and 1. The higher the value of momentum, the more the stability of optimization increases. The updating rule is given by $\theta_{t+1}=\theta_t + \Delta\theta_t$ and $\Delta\theta_t=\alpha\Delta\theta_{t-1}-\eta\nabla C(\theta_{t-1})$, where α is the momentum and η is the learning rate.

The downside of the momentum is that it improves the stability and speeds up the process to reach convergence, but it must be applied by carefully examining the nature of the problem to optimize; otherwise, it results in poor performance. If the value of the momentum is not correctly set, it may lead to overfitting. Sometimes it makes the weights move in the same direction though the gradient is small and stuck in local minima. The following code and Figure 9-8 demonstrate this optimizer:

```
import numpy as npy
import matplotlib.pyplot as mplt
#from mpl_toolkits.mplot3d import Axes3D

# function to be optimized
def f(x,y):
  return x**2.0 + y**2.0

# partial derivatives of the function based on x
def df_dx(x):
  return x * 2.0

# computation of gradient descent
def Mtum(f, df_dx, xyval, num_itr, stp_sz, mtum):
  # record solutions
  solns, sol_scores = list(), list()
  # set the initial position
  soln = xyval[:, 0] + npy.random.rand(len(xyval)) * (xyval
  [:, 1] - xyval[:, 0])
  # initialize the variation to 0.0
  variation = 0.0
  for i in range(num_itr):
```

```
    # compute gradient
    grad = df_dx(soln)
    # compute the current variation
    new_var = stp_sz * grad + mtum * variation
    # next step
    soln = soln - new_var
    # update the variation
    variation = new_var
    # evaluate the solution
    soln_eval = f(soln[0],soln[1] if len(soln) > 1 else 0)
    # save solution
    solns.append(soln.copy())
    sol_scores.append(soln_eval)
    # display the progress
    print(f"\nIteration: {i}, solution :{soln}, evaluation of
    solution: {soln_eval}")
  return [soln, npy.array(solns), sol_scores]

# set the seed value for random number generator
npy.random.seed(7)
# set limit for input
xyval = npy.asarray([[-1.0, 1.0]])
# set the number of iterations
num_itr = 30
# set step size
stp_sz = 0.1
# set momentum
mtum = 0.3
# call momentum function
best_sol, solns, sol_scores = Mtum(f, df_dx, xyval, num_itr,
stp_sz, mtum)
```

```
# set the 3D plot visualization
x = npy.linspace(xyval[0, 0], xyval[0, 1], 100)
y = npy.linspace(-1.0, 1.0, 100)
X, Y = npy.meshgrid(x, y)
Z = f(X, Y)

# Plot the trajectory path
fig1 = mplt.figure()
grph = fig1.add_subplot(111, projection='3d')
grph.plot_surface(X, Y, Z, cmap='viridis', alpha=0.5)
grph.scatter(best_sol[0], best_sol[1] if len(best_sol) > 1 else
0, f(best_sol[0], best_sol[1] if len(best_sol) > 1 else 0),
    color='red', label='Best solution')
grph.plot([pt[0] for pt in solns],
    [pt[1] if len(pt) > 1 else 0  for pt in solns], sol_scores,
    color='blue', label='Trajectory from initial point')
grph.set_xlabel('X')
grph.set_ylabel('Y')
grph.legend()

# visualize the plot
mplt.show()
```

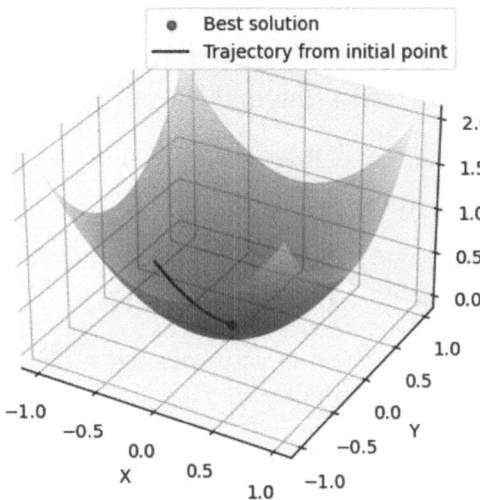

Figure 9-8. *Demonstration of momentum*

9.10 Nesterov Momentum

This type is also known as Nesterov accelerated gradient (NAG); it is an extended version of a traditional momentum optimizer. Momentum itself accelerates the variants of gradient descents using the past gradients to update the weights. Hence, it moves faster over the saddle points and flat spaces and converges quickly. This Nesterov momentum also follows the similar acceleration strategy with a minor modification in the weight updates. It calculates the next iteration gradients in advance in the direction of momentum to correct itself if it is moving toward suboptimal direction. It is more advantageous than momentum because sometimes the acceleration in momentum optimizer makes it skip or overshoot the minima in the valleys, but the Nesterov momentum slows down the search when nearing the minima at valleys or basins to lessen the probability of skipping or overshooting it. It works as follows:

1. The projection position at iteration t+1 of the
   solution is computed using the change at t as
   position(t+1) = x(t) + (mmtm * chg(t)).

2. Compute gradient of this new projection position as
   grdt(t+1) = f '(position(t+1)).

3. Compute change at iteration t+1 as chg(t+1) =
   (mmtm * chg(t)) – (stp_size * grdt (t+1))

4. Finally, compute the new value of each variable as
   x(t+1) = x(t) + chg (t+1)

The following code and Figure 9-9 demonstrate this optimizer:

```
import numpy as npy
import matplotlib.pyplot as mplt

# function to be optimized
def f(x, y):
  return x**2.0 + y**2.0

# partial derivatives of the function based on x and y
def df_dx_dy(x, y):
  return npy.asarray([x * 2.0, y * 2.0])

# nesterov momentum implementation
def NestMtum(f, df_dx_dy, xyval, num_itr, stp_sz, mtum):
  # record solutions
  solns = list()
  # record solution scores
  sol_scores = []
  sol_traj = []
  # set an initial position
  soln = xyval[:, 0] + npy.random.rand(len(xyval)) * (xyval
  [:, 1] - xyval[:, 0])
```

```
# initialize the variations for each variable
variation = [0.0 for _ in range(xyval.shape[0])]
for itr in range(num_itr):
  # compute solution
  sol_projtd = [soln[i] + mtum * variation[i] for i in
  range(soln.shape[0])]
  # compute the gradient
  grad = df_dx_dy(sol_projtd[0], sol_projtd[1])
  # find new solution for eac variable
  new_soln = list()
  for i in range(soln.shape[0]):
    # compute the variation
    variation[i] = (mtum * variation[i]) - stp_sz * grad[i]
    # compute new position for the variable
    value = soln[i] + variation[i]
    # save the variable
    new_soln.append(value)
  # save new solution
  soln = npy.asarray(new_soln)
  solns.append(soln)
  # evaluate the solution
  soln_eval = f(soln[0], soln[1])
  # display the progress
  print(f"\nIteration: {itr}, solution :{soln}, evaluation of
  solution: {soln_eval}")
  sol_score = f(soln[0], soln[1])
  sol_scores.append(sol_score)
  sol_traj.append(soln.copy())
 return soln, sol_scores, sol_traj

# set the seed value for random number generator
npy.random.seed(5)
```

```
# set limit for input
xyval = npy.asarray([[-1.0, 1.0], [-1.0, 1.0]])
# set the number of iterations
num_itr = 50
# set step size
stp_sz = 0.01
# set momentum
mtum = 0.8
# Call nesterov momentum function
solns, sol_scores, sol_traj = NestMtum(f, df_dx_dy, xyval,
num_itr, stp_sz, mtum)

# plot 3D visualization
x = npy.linspace(xyval[0, 0], xyval[0, 1], 100)
y = npy.linspace(xyval[1, 0], xyval[1, 1], 100)
X, Y = npy.meshgrid(x, y)
Z = f(X, Y)

solns = npy.asarray(solns)

fig1 = mplt.figure()
grph = fig1.add_subplot(111, projection='3d')
grph.plot_surface(X, Y, Z, cmap='viridis', alpha=0.5)
grph.scatter(solns[0], solns[1], f(solns[0], solns[1]),
    color='red', label='Best solution')
grph.plot([point[0] for point in sol_traj],
    [point[1] for point in sol_traj], sol_scores,
    color='blue', label='Trajectory from initial point')
grph.set_xlabel('X')
grph.set_ylabel('Y')
grph.legend()

# visualize the plot
mplt.show()
```

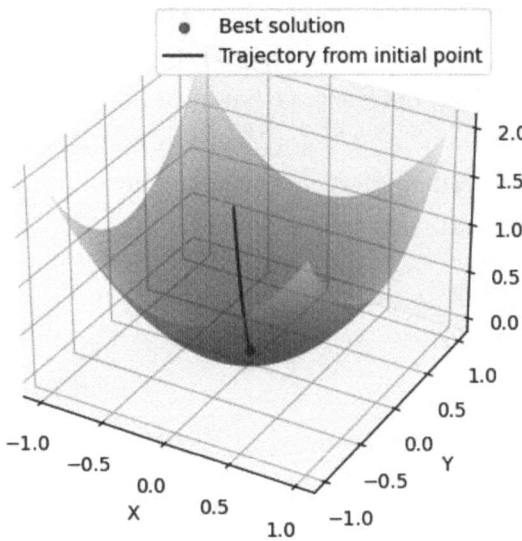

Figure 9-9. *Demonstration of Nesterov momentum*

9.11 Adam

This type is said to be an adaptive moment estimation optimizer, which computes the adaptive learning rate for each parameter involved in learning over every iteration. It also employs bias correction at the earlier iterations to balance the initialization bias. It is an amalgamation momentum and RMSProp for determining the parameters. This type is best for nonconvex functions and noisy functions with sparse gradients.

The hyperparameters used in Adam are α, which is the step size; β_1, which is the decay rate of momentum; β_2, which is the decay rate of squared gradients; and ϵ, which is the small value to avoid divide by zero and generally is set to $1e^{-8}$.

The momentum is given by $m_t = \beta_1 * m_{t-1} + (1-\beta_1)*\text{grad}_t$.

The root mean squared gradient is given by $\upsilon_t = \beta_2 * \upsilon_{t-1} + (1-\beta_2)*\text{grad}_t^2$.

The updation is given by $\theta = \theta - ((\alpha * m_t) / \sqrt{(\upsilon_t + \epsilon)})$.

The Adam optimizer requires very little memory and is suitable for roaming objectives problems, large datasets, more parameters, etc.

The downside of the optimizer is that its performance is highly hyperparameter sensitive, and it requires more computational overhead when compared to gradient descent variants. The following code and Figure 9-10 demonstrate this optimizer:

```python
import numpy as npy
import matplotlib.pyplot as mplt

# function to be optimized
def f(x, y):
  return x ** 2.0 + y ** 2.0

# partial derivatives of the function based on x and y
def df_dx_dy(x, y):
  return npy.array([2.0 * x, 2.0 * y])

# Adam implementation
def ADAM(f, df_dx_dy, xyval, num_itr,alfa, bta1, bta2,
eps=1e-8):
  # set an initial position
  x = xyval[:, 0] + npy.random.rand(len(xyval)) *
  (xyval[:, 1] - xyval[:, 0])
  sol_scores = []
  sol_traj = []

  # moment initialization
  m1 = npy.zeros(xyval.shape[0])
  m2 = npy.zeros(xyval.shape[0])

  for itr in range(num_itr):
    # compute gradient
    grad = df_dx_dy(x[0], x[1])
```

```
    # find solution for each variable one at a time
    for i in range(x.shape[0]):
      m1[i] = bta1 * m1[i] + (1.0 - bta1) * grad[i]
      m2[i] = bta2 * m2[i] + (1.0 - bta2) * grad[i] ** 2
      m1cap = m1[i] / (1.0 - bta1 ** (itr + 1))
      m2cap = m2[i] / (1.0 - bta2 ** (itr + 1))

      x[i] = x[i] - alfa * m1cap / (npy.sqrt(m2cap) + eps)

    # Evaluate solution
    sol_score = f(x[0], x[1])
    sol_scores.append(sol_score)
    sol_traj.append(x.copy())

  return x, sol_scores, sol_traj

# set limit for input
xyval = npy.array([[-1.0, 1.0], [-1.0, 1.0]])
# set the number of iterations
num_itr = 60
# Set step size
alfa = 0.02
# Set factor beta1 for avg. gradient
bta1 = 0.8
# Set factor beta2 for avg. squared gradient
bta2 = 0.999

# call Adam function
best_sol, sol_scores, sol_traj = ADAM(f,df_dx_dy, xyval,
num_itr,alfa, bta1, bta2, eps=1e-8)

# plot 3D visualization
x = npy.linspace(xyval[0, 0], xyval[0, 1], 100)
y = npy.linspace(xyval[1, 0], xyval[1, 1], 100)
```

```
X, Y = npy.meshgrid(x, y)
Z = f(X, Y)

# Plot the trajectory path
fig1 = mplt.figure()
grph = fig1.add_subplot(111, projection='3d')
grph.plot_surface(X, Y, Z, cmap='viridis', alpha=0.5)
grph.scatter(best_sol[0], best_sol[1], f(best_sol[0],
best_sol[1]), color='red', label='Best solution')
grph.plot([point[0] for point in sol_traj],
    [point[1] for point in sol_traj], sol_scores,
    color='blue', label='Trajectory from initial point')
grph.set_xlabel('X')
grph.set_ylabel('Y')
grph.legend()

# Show the plot
mplt.show()
```

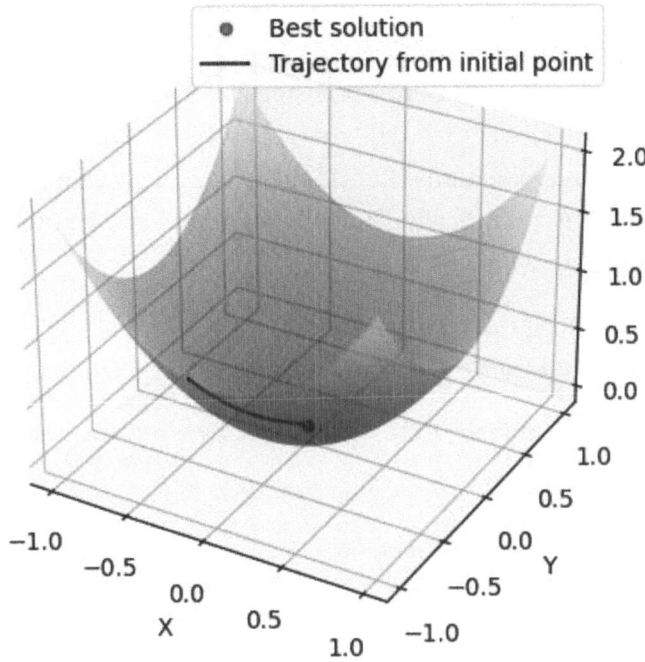

Figure 9-10. Demonstration of the Adam optimizer

9.12 Adamax

This type is the extended version of the Adam optimizer. Adam updates the weights in porportion to the square of the past history of gradients, whereas Adamax updates with a maximum of past gradients. This optimizer automatically adjusts the learning rate for all parameters involved in the learning model. The moment vector and the max of exponentially weighted value for each parameter are denoted as m and v. The gradient is given by $grad(t) = f'(x(t-1))$. The moment vector updation using gradient and hyper parameter β_1 is given by $mv(t) = \beta_1 * mv(t-1) + (1-\beta_1) *grad(t)$. The exponentially weighted infinity norm using the β_2 hyperparameter is given by $ev(t) = \max(\beta_2 * ev(t-1), abs(grad(t)))$. Now the updating takes place with the parameter updation and general updation rule as follows:

287

stp_size(t) = α / (1-β$_1$(t)), Δ(t) = mv(t) / ev(t),
x(t) = x(t-1) – stp_size(t) * Δ (t)

Finally, with the substitution of the step size and delta, the updation becomes x(t) = x(t-1) – α / (1-β$_1$(t)) * mv(t) / ev(t).

Adamax is for applications such as speech or music whose noise conditions change dynamically. The following code and Figure 9-11 illustrate this optimizer:

```python
import numpy as npy
import matplotlib.pyplot as mplt

# function to be optimized
def f(x, y):
  return x**2.0 + y**2.0

# partial derivatives of the function based on x and y
def df_dx_dy(x, y):
  return npy.asarray([x * 2.0, y * 2.0])

# Adamax implementation
def adMax(f, df_dx_dy, xyval, num_itr, alfa, bta1, bta2):
  solns = list()
  # set an initial position
  x = xyval[:, 0] + npy.random.rand(len(xyval)) * (xyval
  [:, 1] - xyval[:, 0])
  sol_scores = []
  sol_traj = []
  # moment initialization
  m1 = [0.0 for _ in range(xyval.shape[0])]
  m2 = [0.0 for _ in range(xyval.shape[0])]
  for itr in range(num_itr):
    # compute gradient
    grad = df_dx_dy(x[0], x[1])
    # find solution for each variable one at a time
```

```
    for i in range(x.shape[0]):
      m1[i] = bta1 * m1[i] + (1.0 - bta1) * grad[i]
      m2[i] = max(bta2 * m2[i], abs(grad[i]))
      stp_sz = alfa / (1.0 - bta1**(itr+1))
      dta = m1[i] / m2[i]
      x[i] = x[i] - stp_sz * dta
    sol_score = f(x[0], x[1])
    sol_scores.append(sol_score)
    sol_traj.append(x.copy())
    # display the progress
    print(f"\nIteration: {itr}, solution :{x}, score of
    solution: {sol_score}")
  return x, sol_scores, sol_traj

# set the seed value for random number generator
npy.random.seed(1)
# set limit for input
xyval = npy.asarray([[-1.0, 1.0], [-1.0, 1.0]])
# set the number of iterations
num_itr = 60
# set the steps size
alfa = 0.02
# Set factor beta1 for avg. gradient
bta1 = 0.8
# Set factor beta2 for avg. squared gradient
bta2 = 0.99
# call adamax function
best_sol, sol_scores, sol_traj = adMax(f, df_dx_dy, xyval,
num_itr, alfa, bta1, bta2)

# plot 3D visualization
x = npy.linspace(xyval[0, 0], xyval[0, 1], 100)
```

```
y = npy.linspace(xyval[1, 0], xyval[1, 1], 100)
X, Y = npy.meshgrid(x, y)
Z = f(X, Y)

# Plot the trajectory path
fig1 = mplt.figure()
grph = fig1.add_subplot(111, projection='3d')
grph.plot_surface(X, Y, Z, cmap='viridis', alpha=0.5)
grph.scatter(best_sol[0], best_sol[1], f(best_sol[0],
best_sol[1]),
    color='red', label='Best solution')
grph.plot([point[0] for point in sol_traj],
    [point[1] for point in sol_traj], sol_scores,
    color='blue', label='Trajectory from initial point')
grph.set_xlabel('X')
grph.set_ylabel('Y')
grph.legend()

# Visualize the plot
mplt.show()
```

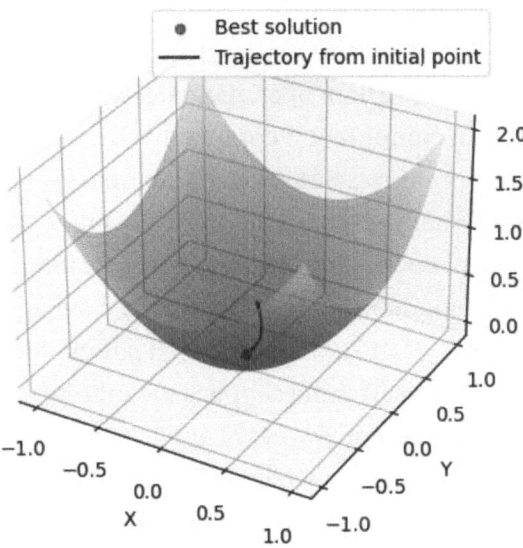

Figure 9-11. *Demonstration of the Adamax optimizer*

9.13 SMORMS3

Squared mean over root mean squared cubed (SMORMS3) is a variant of RMSProp where the squared gradient is replaced with the cube of a squared gradient. Unlike RMSProp computing the average of squared gradients, SMORMS3 computes the cube root of the moving average of the cube of squared gradients, and it helps in preventing the learning rate to decrease quickly. It also helps in preventing the learning rate from becoming too small and slows down the optimization process. In addition, it has a damping factor like RMSProp, which avoids the learning rate from becoming too large.

This is best for the functions that have a high variance in gradient during the optimization process. It offers better performance when the dataset is large and has higher dimensions and steady performance though the data has noisy gradients. The benefit of this type is achieved at the cost of computation expensive requirement. Try coding this by yourself as it requires minor modification in the RMSProp code.

9.14 Summary

Gradient descent and the variant of gradient descent optimizers are simple optimizers but slow in computation comparatively. Adagrad updates the learning rate and and achieves better computational speed. RMSProp and Adadelta are alternatives that use a moving average of gradients and avoid the monotonical decreasing learning rate. Momentum and Nesterov momentum adds the momentum to speed up the convergence process. The Adam optimizer finds the updates for each parameter in each iteration and applies bias correction. Adamax applies the infinite norm of the moving average of the gradient. Sworms3 applies the cube of the squared gradient. Each optimizer has its own benefits. Based on the nature of the problem, the optimizer can be applied to get better results.

Index

A

Activation function, 180
Adadelta, 18, 292
Adaptive gradient, 263
Adaptive gradient descent
    (Adagrad), 18
Additive hierarchical
    clustering, 112
Agglomerative clustering, 112
AlexNet, 216
ANNs, *see* Artificial neural
    networks (ANNs)
APIs, *see* Application programming
    interface (APIs)
Application programming interface
    (APIs), 21, 53
Apriori algorithm, 122, 137
Artificial neural networks (ANNs), 11
    activation function, 180, 181,
        183, 184
    architecture, 179
    communication structure,
        human brain, 178, 179
    definition, 177
    feedback networks, 203–206
    feed forward ANNs, 191

loss function, 185–190
synapse/synaptic junction, 177
types, 191
unsupervised, 207, 208
Autoencoders, 13, 208

B

Back propagation, 180
Backward propagation through
    time (BPTT), 234
Bagging and boosting, 95
Batch gradient descent optimizer, 17
Binary class loss function, 186
Binary step activation function, 180
BPTT, *see* Backward propagation
    through time (BPTT)

C

Classification
    case study, agriculture, 97,
        98, 100–106
    decision trees, 94
    definition, 92
    logistic regression, 93
    naïve bayes, 94, 95